Positive Harmlessness in Practice

Enough for Us All
Volume Two

Dorothy I. Riddle

Dorothy I Riddle

authorHOUSE®

AuthorHouse™
1663 Liberty Drive
Bloomington, IN 47403
www.authorhouse.com
Phone: 1-800-839-8640

© 2010 Dorothy I. Riddle. All rights reserved.

No part of this book may be reproduced, stored in a retrieval system, or transmitted by any means without the written permission of the author.

First published by AuthorHouse 6/24/2010

ISBN: 978-1-4520-3633-5 (e)
ISBN: 978-1-4520-3632-8 (hc)
ISBN: 978-1-4520-3631-1 (sc)

Library of Congress Control Number: 2010902808

Printed in the United States of America
Bloomington, Indiana

This book is printed on acid-free paper.

Enough for Us All is dedicated to my mother, Katharine (Kittu) Riddle, whose loving energy and inquiring spirit have brought joy to me and to hundreds of others all around the world.

It is from my mother that I first learned the meaning of abundance, as well as the worth and dignity of all beings, and began to see beyond the confines of our planetary life and to question the assumptions that seem to hold us captive.

Contents

Preface . . . i

Part One
Understanding Harmlessness

1 **The Concept of Harm** . . . 1
   The root causes of harmfulness . . . 2
   The pervasiveness of harm . . . 4
   International understanding of harm . . . 6
   Harming by commission . . . 12
   Harming by omission . . . 16
   Healing the effects of harm . . . 18

2 **Positive Harmlessness as Our Core Value** . . . 23
   Definitions of harmlessness . . . 23
   Harmlessness and the seven principles . . . 29
   Harmlessness in thought . . . 38
   Harmlessness in word . . . 40
   Harmlessness in action . . . 42
   Valuing positive harmlessness . . . 45

3 **Harmlessness and the Butterfly Shift** . . . 47
   Becoming conscious of the need to change . . . 48
   Our incentive to change . . . 53
   Making a change permanent . . . 55
   Introducing the Butterfly Shift dynamic . . . 58
   The Butterfly Shift . . . 61
   Leveraging the Butterfly Shift . . . 63

Part Two
Immersion in Harmlessness— The Butterfly Shift

4 **Managing Our Focus** . . . 67

	Focusing our attention	68
	Mindfulness and mindlessness	73
	Who we notice	75
	What we notice	79
	Leveraging our ability to focus	82
5	**Noticing — Step One**	**85**
	Ways to improve your noticing skills	85
	Focus and the type of shift	87
	Choosing your focus	87
	Ensuring the success of Step One	90
6	**Leveraging Emotions**	**91**
	The importance of emotions	91
	Developing our emotional muscle	94
	Emotional contagion	103
	The power of emotions	105
	The heart and the brain	106
	Leveraging specific emotions	107
	Maximizing benefits from emotions	111
7	**Feeling — Step Two**	**113**
	Ways to improve your feeling expression	113
	Emotion and the type of shift	115
	Choosing your feeling expression	118
	Ensuring the success of Step Two	118
8	**Reviewing Our Action Options**	**121**
	Becoming noticed	121
	Addressing a person's needs	123
	Actions and values	124
	Cultural filters	125
	Feedback and harmlessness	130
	Actions and the Joyous Shift	133
	Detachment and the Butterfly Shift	134
	Facilitating the mini-immersion	136
9	**Acting — Step Three**	**137**
	Ways to improve your action effectiveness	137
	Action and the type of shift	140
	Choosing the type of shift	141

Ensuring the success of Step Three   142

Part Three
Ensuring Ongoing Harmlessness

10 Maturing Into Harmlessness **145**
    The focus of our developmental models   146
    Our developmental context   149
    Our starting point   151
    Maturing in relation to others   154
    Our developmental process   156
    Harmlessness and moral development   161
    An alternate view of maturation and purpose   162
    Developing a model that supports harmlessness   167

11 Our Maturational Opportunity **169**
    Redefining our maturational goal   169
    Maturing in self-discipline   175
    Maturing in responsibility   177
    Maturing in decision making   180
    Maturing in complexity   183
    Maturing in nurturance   185
    Maturing in goodwill   187
    Maturing in compassion   190
    Ensuring the habit of harmlessness   192

12 Developing an Ethic of Harmlessness **193**
    Reclaiming harmlessness as a strength   194
    From reciprocity to harmlessness   198
    Bridging the gap between theory and practice   204
    Measuring our engagement with harmlessness   208
    Gender harmlessness as a litmus test   215
    Practicing harmlessness   222

Appendices:
    A. Universal Declaration of Human Rights   225
    B. *Ending Violence Against Women*   231
    C. Harmlessness Scale™: Questions & Scoring   235

Notes	245
References	269
Index	277
List of Exercises	287

Preface

The *Enough for Us All* series was inspired by the architect and futurist Buckminster Fuller's 1980 assertion that we do now have the capacity to "take care of everybody at a higher standard of living than any have ever known." The three volumes are designed to help us recognize and embrace our abundant and joyous cosmic reality. They explore both the personal and the societal aspects of shifting from a preoccupation with scarcity to participation in a collaborative process. In each volume, there are many practical exercises to help us apply the concepts in daily life.

This second volume, *Positive Harmlessness in Practice*, focuses on the meaning of positive, or proactive, harmlessness. Although harmlessness is mandated by every spiritual tradition, it is challenging to actually implement. Exploring the concept of harmlessness makes it clear that we have no collective experience of harmlessness because our habits of harm are so pervasive. It is difficult to embed harmlessness as our ethic without consciously experiencing it. To build our "harmlessness muscle," the book details a pragmatic three-step daily practice—a Butterfly Shift—that we can easily and quickly undertake in order to have small, mini-immersion experiences of harmlessness.

But a lasting shift to an ethic of harmlessness requires more than periodic practice. Given that we are interconnected, interdependent, energetic beings, we need to re-examine how we focus our energy so that we empower, rather than harm, ourselves and others. A proposed Harmlessness Scale™ helps us identify our habitual ways of behaving so that we can shift to automatic patterns of harmlessness. In closing, the book challenges us, as a human family, to demonstrate our commitment to an ethic of harmlessness by eliminating the rampant violence against women that currently constitutes, worldwide, the single greatest human rights violation.

Principles That Underlie Our Reality	
Interconnectivity	We are all interconnected energy waves.
Participation	We create our own reality.
Nonlinearity	Our experience is fundamentally nonlinear.
Nonduality	Our reality is complex and non-dualistic.

Principles That Govern How We Coexist	
Interdependence	We are part of an interdependent community of life.
Adaptability	We survive because of our ability to adapt and collaborate.
Cooperation	We evolve through symbiosis and cooperation.

Volume One, *Principles of Abundance for the Cosmic Citizen*, explores seven principles (listed above)—four that underlie how our reality operates, and three that govern our existence in the cosmos. Each discussion identifies and explores the limiting beliefs that we have acquired over the centuries. Until we are clear about who we actually are and our intended relationship with the rest of life, we are not in a position to actualize our potential and shift from fear to joy as our basic motivation.

Volume Three, *Moving Beyond Duality*, exposes the illusion of duality that underlies our fear of scarcity and helps us learn to live joyously and interdependently.

I would like to thank all who have helped me in my journey of exploration and questioning. That journey has included the privilege of living and working in over 75 developing countries, establishing the first degree-granting women's studies program in 1971 at City University of New York, undertaking the initial research on homophobia in the early 1970s, and studying and working over the years with the School for Esoteric Studies. In celebrating the finalization of these volumes, I would like to thank my various *Writer's Digest* instructors, particularly Carolyn Walker who has provided editing commentary and much valued encouragement. I would also like to thank my partner, Valerie Ward, for her continuing and invaluable support for my creative process, as well as Barbara Austin for her help in birthing this series, Bernadette Richards and Miguel Malagreca for their critical input, and other friends who have read and commented on earlier versions.

May we learn together that joy is the keynote of our universe and that there is indeed enough for us all.

Part One

Understanding Harmlessness

ONE

The Concept of Harm

*If you can, help others;
if you cannot do that, at least do not harm them.*

— Dalai Lama

All religious traditions share a commitment to harmlessness, however defined. So do professional codes of conduct, which include phrases like that attributed to the Hippocratic Oath: "First, do no harm."

Why do we tolerate harm? Why is there so much violence? It is clear that we cannot rely on spiritual beliefs to prevent it. Part of the answer lies in our distorted view of who we are and how our universe operates, which encourages us to focus on self-interest and to live from a fear-based myth of scarcity.

The first volume in this *Enough for Us All* series, titled *Principles of Abundance for the Cosmic Citizen*, explored the meaning of "enough," the benefits and responsibilities of being a cosmic citizen, and seven principles by which our cosmos operates (which are listed in the Preface of this volume). We will review those principles in the next chapter as they relate to harmlessness. In exploring the concept of harm, the principles that concern us most are that we are indeed all part of the same cosmic energy field (the Principle of Interconnectivity) and that we have evolved primarily through cooperation and networking (the Principle of Cooperation).

2 Enough for Us All: Positive Harmlessness

We have been shaped over the past 300 years by the mechanistic, deterministic worldview that emerged from Newtonian physics as well as by a belief that our evolutionary history was grounded in violence and competition ("survival of the fittest"). The findings of quantum physics reveal quite a different picture and one that can help us shed our mistaken identities as violent beings. But first, if we are to shift from harmfulness to harmlessness, we need to understand harm and why we indulge in it.

The Root Causes of Harmfulness

Why do people choose to behave in a harmful or violent manner? We may act harmfully out of fear, particularly a fear of scarcity or of loss. If we see life as a win-lose competition for scarce resources, then it is easy to justify any action that will make us the winner. This is an "I'll get them before they get me" mentality.

Or we may act harmfully out of an ignorance of alternatives. If all that is modeled for us is the option of violence, then it would come as no surprise that we choose violence. Our violence may be a matter of unconscious habits or "the way it's always been done"—as in hazing, corporal punishment, or forced sex. We may not have developed a moral compass of our own.

Or we may act harmfully because of our own sense of entitlement to have what we want when we want it. In this scenario, anyone who interferes is expendable. We explored this entitlement dynamic in some detail in *Principles of Abundance for the Cosmic Citizen*. In essence, here we choose to make ourselves feel good at the expense of others.

Or there is the issue of choice and control. Causing harm is a control strategy, not an out-of-control occurrence. Some people make choices to dissipate tension and frustration through violence rather than accepting responsibility to work through the discomfort without harming those around them.

> **Exercise: Motivation Regarding Harm**
>
> Reflect back on the past week and identify the following:
>
> 1. I had an opportunity to speak negatively about someone, and I consciously refrained because:
>
> 2. I had an opportunity to act aggressively towards someone, and I willfully refrained because:
>
> 3. I could feel myself becoming judgmental about another person, and I purposely shifted to compassion because:
>
> How did you feel as a result of the types of actions described above?

The motivation of choice raises the matter of intention. Part of the definition of harm refers to hurt that is deliberately inflicted. Intention is tricky because we can have good intentions and yet harm others without realizing what the consequences of our actions will be.

Finally and most fundamentally, we may do harm to others in denial of our actual interconnected reality, substituting instead an "us-them" mentality that includes objectifying the other person or group. When we remember that the "other" is in a very real sense a part of ourselves—the Principle of

Interconnectivity—then it is not so easy to humiliate or degrade or physically hurt them.

The Pervasiveness of Harm

Surrounded as we are by messages about the importance of harmlessness, one would think that harming another person would seldom occur. Unfortunately, we know better. Harmfulness is ubiquitous. We kill millions in wars over religious differences and territorial greed, leaving millions more scarred for life. Murders, assaults, sexual and domestic violence, and other examples of violence between humans are rampant, to say nothing of the mistreatment of animals, species extinction due to human initiatives, or damage to the environment. And, though caused unintentionally, harmful illnesses stemming from misdiagnosis or inappropriate treatment are the third leading cause of death in the United States, after heart disease and cancer.[1] The list seems endless.

Rather than improving, our rate of violence appears to be escalating. In fact, the World Health Organization (WHO) has declared that "the 20th century was one of the most violent periods in human history. An estimated 191 million people lost their lives directly or indirectly as a result of conflict, over half of whom were civilians."[2]

Our callousness seems to be intensifying as the amount of violence in the media, music and music videos, and video games increases exponentially. Violence generates an adrenalin rush that can fill us with excitement. How to generate a similar sense of excitement without the accompanying violence is not clear.

Recent concern regarding violence has centered around not only its quantity but also the increasing proportion of gratuitous violence—that is, violence that does not contribute to understanding a plot or to advancing a storyline. This gratuitous violence includes a rising percentage of violence depicted without moral consequences or posed as the normal solution to interpersonal difficulties, as well as the persistent linking of violence with sexual and graphic sadist imagery.[3]

Over the years there has been debate about whether watching or listening to violence has an effect on our behavior. The evidence now appears conclusive that indeed there are negative consequences. Research indicates that viewing or listening to or acting out violence reinforces a belief that violence is the best way to resolve conflicting interests. It also creates an image of the world as a dangerous place where one must be on guard.[3]

Exercise: Violence in Media

Think back over the past 24 hours. Did you "absorb" any violent messages from newscasts, newspapers, magazines, movies, music, books, etc?

If "no," did you "neutralize" incoming messages or did you avoid them altogether?

If "yes," was the impact offset by messages about the inappropriateness of violence or your own internal dialogue about the need to end violence?

What would you do differently if you committed to zero tolerance for violence?

In fact, the American Academy of Pediatrics issued the following policy statement in October 2009: "Exposure to violence in media, including television, movies, music, and video games, represents a significant risk to the health of children and adolescents. Extensive research evidence indicates that media violence can contribute to aggressive behavior, desensitization to violence, nightmares, and fear of being harmed."[4]

In 2002, WHO released the *World Report on Violence and Health*, which covered a range of individual and collective violence (such as war), including child abuse and maltreatment, elder abuse, intimate partner violence, self-directed violence, sexual violence, and youth violence. Violence and related injuries account for over five million deaths annually, making it the third leading cause of death worldwide after heart disease and stroke.[5] Declaring violence as a leading worldwide public health problem, WHO launched a Global Campaign for Violence Prevention in 2003, which is still ongoing.[6]

In 2006, the United Nations (UN) issued a 140-page report, "In-Depth Study of All Forms of Violence Against Women," in which, for the first time, violence against women and girls was confirmed to be a human rights violation.[7] The UN Secretary-General has identified violence against women and girls, in particular, as the most widespread human rights violation in the world and has launched the UNiTE campaign to end violence against women.[8]

International Understanding of Harm

Before we examine what harmfulness in general means in our everyday lives, let's take a few moments to trace our collective definition of what constitutes harm. We have already seen that two of our major intergovernmental organizations—the World Health Organization and the United Nations—are engaged with this issue.

As we have gradually come to realize that we are all interconnected, we have been shifting what we define as acceptable human behavior. Less than 200 years ago, for example, slavery was accepted as an economic necessity; today it is considered reprehensible (though it still occurs). The children of those with fewer economic resources used to be unschooled and expected to earn a living from an early age. Now the second UN Millennium Development Goal is universal primary education,[9] and child labor laws are in effect in many countries. Religious wars and the accompanying intolerance of differences used to be common, while now we have a growing global concern with inclusiveness and understanding as reflected in movements like the United Religions Initiative.[10] Many other examples of the growing recognition of "unity in diversity" can be found at *www.servicegrowth.net* by clicking on Spirituality in Practice / Global Initiatives.

Our commitment to a shared understanding of "harm," though, is relatively new, especially one that is applied equally to all persons. That understanding has expanded from a focus on only physical hurt to include "mental, moral, or spiritual injury."[11] In earlier times of dictatorial monarchies, there were widely-accepted differences in the standards of justice applied by social class. It is really only since World War II, with the formation of the United Nations and the launching of the Nuremberg Trials, that we have begun to formulate a collective sense of harm and human rights, culminating in the concept of *crimes against humanity*.

We may accept that harming others harms ourselves (based on the Principle of Interconnectivity), but what exactly do we consider collectively harmful? The concept of crimes against humanity began with an official outcry against the Armenian genocide, issued on May 24, 1915. This joint statement by Britain, France, and Russia was the first such charge against another government: "In view of these new crimes of

Turkey against humanity and civilization, the Allied Governments announce publicly to the Sublime Porte that they will hold personally responsible for these crimes all members of the Ottoman Government, as well as those of their agents who are implicated in such massacres."[12]

> ### Exercise: Addressing Harm
>
> Think back over the last week and identify examples of each type of harm that you have inflicted, experienced, or witnessed:
>
> 1. Physical
> 2. Mental
> 3. Emotional
>
> If you were the perpetrator, why? How could you have kept the harm from happening?
>
> If you were the recipient, how could you have protected yourself and helped to keep it from happening again?
>
> If you were the witness, did you intervene? Why and how, or why not?

Thirty years later, we faced the inhumane torture and eradication of millions of people during the Holocaust and wondered how it could have occurred. The Nuremberg Trials, which started in 1945 to address the atrocities of the Third Reich and its collaborators, was the first international criminal tribunal ever to be convened.[13] To hold such a tribunal, there first had to be an agreement on what was considered to be illegal, or harmful, behavior. And so the Nuremberg Principles, listed below, were born.

The Nuremberg Principles

I Any person who commits an act that constitutes a crime under international law is responsible therefore and liable to punishment.

II The fact that internal law does not impose a penalty for an act that constitutes a crime under international law does not relieve the person who committed the act from responsibility under international law.

III The fact that a person who committed an act that constitutes a crime under international law acted as Head of State or responsible government official does not relieve that person from responsibility under international law.

IV The fact that a person acted pursuant to order of Government or of a superior does not relieve the person from responsibility under international law, provided a moral choice was in fact possible.

V Any person charged with a crime under international law has the right to a fair trial on the facts and law.

VI The crimes hereinafter set out are punishable as crimes under international law:

   (a) **Crimes against peace:**

> (i) Planning, preparation, initiation, or waging of a war of aggression or a war in violation of international treaties, agreements, or assurances;
>
> (ii) Participation in a common plan or conspiracy for the accomplishment of any of the acts mentioned under (i).
>
> (b) **War crimes:**
>
> Violations of the laws or customs of war that include, but are not limited to, murder, ill-treatment or deportation of slave labor or for any other purpose of the civilian population of or in occupied territory; murder or ill-treatment of prisoners of war or persons on the Seas, killing of hostages, plunder of public or private property, wanton destruction of cities, towns, or villages, or devastation not justified by military necessity.
>
> (c) **Crimes against humanity:**
>
> Murder, extermination, enslavement, deportation, and other inhumane acts done against any civilian population, or persecutions on political, racial, or religious grounds, when such acts are done or such persecutions are carried on in execution of or in connection with any crime against peace or any war crime.
>
> VII Complicity in the commission of a crime against peace, a war crime, or a crime against humanity as set forth in Principle VI is a crime under international law.

These Nuremberg Principles set out for the first time in human history the types of crimes that are punishable under international law: crimes against peace, war crimes (in essence violations of the Geneva Conventions regarding the treatment of prisoners of war[14]), and crimes against humanity. They also described the conditions under which an individual could be held accountable by being placed on trial for violations of international law. These included acting under the direction of a superior and even if there were no national law forbidding the action.

The Nuremberg Principles were followed by the UN's Universal Declaration of Human Rights of 1948,[15] which provides the moral framework regarding the "basic standards without which people cannot survive and develop in dignity," and in turn underpins international humanitarian law. The Preamble to the UN Universal Declaration of Human Rights states that "recognition of the inherent dignity and of the equal and inalienable rights of all members of the human family is the foundation of freedom, justice, and peace in the world"; and Article 3 asserts that "everyone has the right to life, liberty, and security of person." The UN has further established a number of Conventions that enforce this Universal Declaration of Human Rights, starting with one on preventing genocide in 1951.[16]

The establishment of the International Criminal Court (ICC) at The Hague, under the Rome Statue, in July 2002 was another step forward in a collective commitment to contain and address violence under international humanitarian law.[17] In contrast to the UN International Court of Justice, which focuses on settling disputes between States that are members of the UN, the ICC is a permanent tribunal with the legal authority to prosecute individuals for crimes against humanity, genocide, and war crimes when national courts are unwilling or unable to investigate or prosecute. The Rome Statute, in expanding on the Nuremberg Principles, balances punishment of criminals with provisions for victims to be heard and to receive reparations for their suffering—another step forward for humanity.

The ICC's Rome Statute Explanatory Memorandum specifies that the systematic practices that constitute crimes against humanity "are particularly odious offences in that they constitute a serious attack on human dignity or grave humiliation or a degradation of one or more human beings." This clearly underscores the UN Universal Declaration of Human Rights' commitment to persons' inherent dignity. And we see a beginning shift from an emphasis on only physical harm to damage to the human spirit.

But the practice of harmfulness is more deeply embedded in our thinking and our current social structures than this discussion of physical harm and psychological degradation implies. We not only tolerate violence routinely, but we also glorify it in our superheroes and our rites of passage, especially for young men. And nations engage in "just wars" where "war provides a reason to be violent."[18]

Harming by Commission

Refocusing from the global context to our own individual lives, how do we understand "harm" in our everyday activities? What are the parallels in our lives to our collective international understanding?

Our local and national legal frameworks have historically focused primarily on overt physical actions, in much the same way that the definition of "harm" initially focused on physical hurt. Did we beat or rape or steal from someone? Did we poison a public water supply or place an explosive in public transportation? We now have general agreement that any actions like these are harmful and that perpetrators should be punished.

When we examine the Rome Statute, we see a listing of very specific actions that, by their very nature, typically take place in public arenas during armed conflicts, with plenty of witnesses. In everyday life, though, the lines between harm and harmless blur and become less distinct. Many of our interactions with others are private and not subject to public scrutiny. We can act warm and loving in public and be denigrating and spiteful in private. Does the lack of an audience absolve us?

No, of course not. Only the very young and immature would say that "anything goes as long as you don't get caught." Instead, we rely on ethical principles and values to make choices about our behavior.

Exercise: Identifying Harmful Actions

Think back over the past week and identify three times when you have acted in a harmful manner to yourself or someone else:

1.

2.

3.

Why did you act to cause harm?

What would help you not do it again?

Harkening back to the UN commitment to human dignity, we have learned in recent years that harm to the spirit is as damaging as harm to the physical body. In recognizing this, we have developed the civic concept of harassment, which includes verbally offending or humiliating people and interfering with their ability to learn and work.

Harassment includes several categories of behavior that may occur once or many times: discriminatory behavior, personal harassment, sexual harassment, abuse of authority, poisoned work environment, and bullying.

Discriminatory behavior refers to treating people negatively based on prohibited grounds of discrimination such as race, color, ancestry, political beliefs, religion, age, sex, sexual orientation, marital status, family status, disability, or pardoned criminal conviction.

Personal harassment includes conduct that demeans, belittles, or causes personal humiliation or embarrassment to the victim. Much of what we characterize as psychological or emotional abuse falls into this category.

Sexual harassment refers to any conduct of a sexual nature that might reasonably be expected to cause offence or humiliation or be perceived as placing a condition of a sexual nature on employment, an opportunity for training or promotion, receipt of services, or a contract. Examples of behavior that can constitute sexual harassment include:

- unwanted touching, patting, or leering
- sexual assault
- inquiries or comments about a person's sex life
- telephone calls with sexual overtones
- gender-based insults or jokes causing embarrassment or humiliation

- o repeated unwanted social or sexual invitations
- o inappropriate focus on a person's physical attributes or appearance

While the categories of behavior just outlined could occur anywhere, the following two categories are specific to the workplace. *Abuse of authority* refers to improperly using the power inherent in a position to endanger a person's job, undermine job performance, threaten economic livelihood, or in any way interfere with a person's work life. It involves the exercise of authority to intimidate, threaten, blackmail, or coerce the person.

A *poisoned work environment* is characterized by any activity, not necessarily directed at anyone in particular, that creates a hostile or offensive workplace. Examples of a poisoned work environment include graffiti, sexual, racial or religious insults or jokes, and displays of pornographic or other offensive material.

Bullying is a relative newcomer to the harassment family. It has existed for generations in forms like hazing or shunning, but we are now starting to question their acceptability. Only recently have we recognized the long-term psychological damage from being bullied. Direct bullying includes intimidating or humiliating the person by actions like shoving and poking, choking, punching and kicking, beating, pulling hair, scratching, biting, and pinching.

Indirect bullying, on the other hand, is more insidious and includes unjustified criticism and trivial fault-finding that humiliates the person, especially in front of others. It includes name calling, staring, giggling, and laughing at the person. It also includes precipitating the social isolation of the person by spreading lies and gossip, refusing to socialize with the person, and threatening other people who do wish to socialize with the person. A rapidly-growing recent variation that has resulted in teen suicides is cyber bullying, or the relentless posting of negative messages (as extreme as, "You should die") on Facebook or transmitted via instant messaging.[19]

In a work context, superiors may bully subordinates by setting the subordinate up for failure through creating unrealistic goals or deadlines, or deliberating withholding necessary information and resources; either overloading the subordinate with work or taking all work away (sometimes replacing proper work with demeaning jobs); or increasing responsibility while removing authority.

In each type of harassment described, we see how the recipient is degraded and treated like an object instead of a person. Once we dehumanize someone, we make it easy to feel like "anything goes." What does this say about ourselves if we treat another part of our energetic field with such disdain?

Harming by Omission

The Nuremberg Principle that is not yet well defined in civil society is Principle VII stating that "complicity in the commission of a crime against peace, a war crime, or a crime against humanity" is itself a crime. In practice, this has been interpreted as active support in wrongdoing. How active is "active"? If we know that others are being harmed and we do nothing to intervene, are we also guilty of the crime?

In general, we become complicit in, or accessories to, a crime if we are aware of its occurrence (or planned occurrence), have the ability to report that crime to the proper authorities but fail to do so, and are aware that our inaction will allow the crime to occur or continue. Note that being an accessory to a crime does not necessarily require that we participate or even be present. We can also be an accessory by facilitating the crime or its continuance through our inaction.

Under common law, accessories are generally considered as guilty as the principals.

When we translate this concept into everyday behavior, it implies that we have a responsibility to prevent harm, not just to abstain from deliberately causing it. Let's take an example of observing a young boy being bullied on a school playground out of sight of the teacher on duty. What is our responsibility? We could intervene directly if we felt able, but would that really stop the bullying in the long run? We could walk on by, saying that we have no role on the school ground. We could seek out the teacher on duty, or even the school principal, and report what we have seen with enough detail that the perpetrators could be identified, and then leave the consequences in their hands.

What if the violence is more widespread? What if it has already been documented, as in the crimes against humanity being reviewed by the ICC? Do we have any responsibility to mobilize public opinion in support of the recipients? What about a situation like the Democratic Republic of Congo where, although untold thousands of persons are being killed and mutilated, the UN has withheld the peacekeeping forces that have already been authorized? How do we behave harmlessly in such circumstances? These are not easy questions.

One other issue of omission that is relevant is the position of non-interference. Conscientious objectors, in particular, struggle with the consequences of purposeful noninvolvement when violence is being perpetrated. What if the knowledge that we will not act to stop violence is itself a trigger for more violence? For example, in the Rwandan genocide of 1994, the commander of the UN peace keeping mission, Lt. General Roméo Dallaire, requested additional troops and authorization to intervene in order to bring the genocide to an early end. Instead, he was instructed to restrict activities to evacuating foreign nationals from Rwanda. This left no effective restraint on the violent actions of Hutu militants, with horrendous consequences.

So the issue of harm by omission lies in knowing that violence is being, or is predicted to be, done and deliberately doing nothing. If we are committed to harmlessness, we must be prepared to intervene or speak up when others are being harmed.

> **Exercise: Identifying Harmful Inactions**
>
> Think back over the past week and identify three times when you could have acted to prevent harm to yourself or others:
>
> 1.
>
> 2.
>
> 3.
>
> Why did you choose not to act?
>
> What would help you choose to act to prevent harm in the future?

Healing the Effects of Harm

Research has shown that focusing on revenge simply keeps a wound open and fresh.[20] So how do we reach closure when harm is done and begin to heal so that we can move forward in harmlessness? We now have extensive experience with the Truth and Reconciliation process, for which South Africa has been a model. Similar processes are underway in Argentina, Canada, Chile, El Salvador, Fiji, Ghana, Guatemala, Liberia, Morocco, Panama, Peru, Republic of Korea, Sierra Leone, Solomon Islands, East Timor, and the United States. Such processes are intended to allow societies to heal after major internal conflicts and to prevent historical revisionism of state terrorism and human rights violations.

Experience has taught us that a "Truth and Reconciliation" process of healing has at least four stages. First, we must acknowledge what happened through remembering or reconstruction or testimony—in other words, get all the facts out. This includes hearing from the recipient, as has become part of the ICC proceedings.

In some instances, the harm originally occurred far enough in the past that most or all of the recipients are now dead. Still the acknowledgement is critical "lest we forget" and repeat the horrors. One case in point is the sex slavery (that is, "comfort women" forced to sexually service Japanese soldiers) that was perpetrated during World War II by Japan under Emperor Showa. In December 2000, the Women's International War Crimes Tribunal on Japan's Military Sexual Slavery was held in order to gather testimony from recipients and then, based on the international laws that were in place during World War II, to try groups and individuals for rape or sexual slavery. On December 4, 2001, the Tribunal's over 200 page final judgment was issued in the Hague. The last two paragraphs of the final judgment eloquently express the importance of bearing witness to what happened:

> The Crimes committed against these survivors remain one of the greatest unacknowledged and unremedied injustices of the Second World War. There are no museums, no graves for the unknown "comfort woman," no

education of future generations, and there have been no judgment days for the victims of Japan's military sexual slavery and the rampant sexual violence and brutality that characterized its aggressive war.

Accordingly, through this Judgment, this Tribunal intends to honor all the women victimized by Japan's military sexual slavery system. The Judges recognize the great fortitude and dignity of the survivors who have toiled to survive and reconstruct their shattered lives and who have faced down fear and shame to tell their stories to the world and testify before us. Many of the women who have come forward to fight for justice have died unsung heroes. While the names inscribed in history's page have been, at best, those of the men who commit the crimes or who prosecute them, rather than the women who suffer them, this Judgment bears the names of the survivors who took the stand to tell their stories, and thereby, for four days at least, put wrong on the scaffold and truth on the throne.[21]

Second, reparations are required. At a minimum, this involves a sincere apology by the perpetrator that recognizes what was done and the harm inflicted. To be effective, that apology needs to also indicate why it will never happen again. Often there are also actions required to compensate the victim—money, time, effort. One of the interesting initiatives in the United States is the creation of Youth (or Teen) Courts in 25 states, where youth offenders face a jury of their peers who recommend an appropriate sentence, usually comprised of reparations, community service, and also service in a Youth Court.[22] Recidivism rates are half of those of the traditional juvenile court system.

Third, any systemic issues that allowed the harm to occur need to be addressed so that the victim has some reason to believe that the harm will not recur. In addition to individual perpetrators making promises, this might include education and training of the public on the issues involved, better oversight, or specific accountabilities being put in place.

In Canada, a system of Aboriginal courts is being developed within the formal legal system but with the latitude to proceed in accordance with traditional Aboriginal values. For example, since November 2006 there has been a First Nations Court in British Columbia where all parties to a conflict sit together around a table.[23] After the charge is read and a guilty plea entered, everyone has a chance to participate in the discussion. Then a healing plan, which includes support from all relevant parts of the community, is developed by consensus in order to address the root cause of the criminal act.

Finally, there is the matter of forgiveness. Unresolved bitterness is itself harmful to the victim. Forgiveness is a dynamic that is often misunderstood. The intention is *not* to pardon or pretend that everything is just fine, as though the harm had never occurred. Rather, it is to recognize that we share a common humanity and that we all make mistakes, sometimes terrible ones.

Exercise: Forgiveness—Part 1

Think of a time when you were harmed and you were unwilling to forgive the offender.

Why were you unwilling to forgive?

What impact has holding onto that hurt had on your life?

What would make it possible for you to forgive?

Forgiveness is based on compassion for the perpetrator and a desire to release the damage that criticism and hatred can do to our own spirits. One of the most inspiring descriptions of the dynamic and power of forgiveness is Azim Khamisa's *From Forgiveness to Fulfillment*.[24] The author describes how his son was murdered by a 14-year-old as part of initiation into a youth gang in January 1995 and how he was able to realize that "there were victims at both ends of the gun." Since November 1995, he and the boy's grandfather have been working, through the Tariq Khamisa Foundation, in public schools to teach nonviolence and end kids killing kids. In addition to visiting his son's murderer in prison and forgiving him, the author has petitioned the Governor of California (so far unsuccessfully) for his early parole and offered him a job in the Foundation.

> ### Exercise: Forgiveness—Part 2
>
> Now think of a time when you were harmed and you *were* able to forgive the offender.
>
> What made it possible for you to forgive?
>
> What has been the impact on you?

Until we heal from violence we cannot move forward into positive harmlessness. With healing, there is hope for a harmless future.

TWO

Positive Harmlessness as Our Core Value

Harmlessness is a challenging concept. It starts to almost dissolve when we examine it in too much detail. Harmlessness towards whom? Our family and friends? People from all nations and walks of life? Friendly people? Even violent aggressors? What about towards animals? Towards the Earth and its environment? We quickly blanch at the ramifications.

We grow up knowing that it is wrong to harm others, and yet the boundaries of what is harmful are not clear. We live immersed in a matrix of harm that is almost invisible to us because it is so pervasive. Our society is filled with violence of all types—the most visible types of harm. We have already seen, though, that we can harm others and ourselves not just by what we do but also by failing to act. If we have so much trouble grasping the wide-ranging effects of harm, how can we understand and embrace harmlessness?

Definitions of Harmlessness

What is harmlessness, anyway? What behavior is included? Our physical actions? Our emotional responses? Our comments? Our silence? Our thoughts? How are we supposed to decide what to do or not do?

When we look at how we define harmlessness, what we find is a series of contradictory meanings. They range from negative to positive, from reactive to proactive, from what not to do to actions that we should take. No wonder we are not very clear on the concept!

Innocuousness

At one end of the spectrum, "harmless" is defined as "not able or likely to cause harm."[1] Here we have personality descriptions like mild or unobjectionable or even gullible.[2] This is the meaning we use when we say, "He looks harmless," meaning that he looks innocuous or inoffensive. It includes the concept of not wanting to hurt anyone's feelings or not wanting to offend anyone . . . and so being ineffective. We even find "harmless" referring to being weak in the sense of not having the power or capacity to inflict harm.

Exercise: Redefining Innocuousness

Pick someone you know who is mild-mannered and slow to anger and who you've had an opportunity to observe in a situation that could have become confrontational.

How did the person defuse the potential confrontation?

What do you think was the person's motivation?

On reflection, was the way they behaved easy or difficult?

Is this how we view the concept of harmlessness—as a lack of potency or a lack of ability to be effective? If we reflect more deeply, are we sure we admire harmlessness? At this end of the spectrum, do we equate being harmless with the positive attributes of being kind, gentle, trustworthy? Or does a part of us suspect that being harmless means having the negative traits of being naïve and powerless? Is harmlessness an indication of strength or of weakness?

The tension between "harmless" as a sign of strength or as a sign of weakness is particularly apparent when we refuse to return harm for harm. "Turn the other cheek" is a common Christian admonition from the Sermon on the Mount, but is it actually admired? Or is it considered sissy? Is it easy or difficult to refuse to engage in a violent encounter?

So, at this end of the spectrum, we find harmlessness defined as a deficit or lack—the lack of potency or the lack of the ability to be harmful. From this perspective, violence can actually seem to be a way of proving that we *are* potent or powerful.

Avoiding Harm or Violence

A common way of defining harmlessness in action terms, rather than as a lack or inaction, is as "causing no harm."[3] Here harm is defined as hurt, damage, wrong, or even evil and is assumed to cause pain, loss, or suffering.

Ancient scriptures from the Buddhist, Hindu, and Jain traditions focus on this definition in the concept of *ahimsa*, or "without violence"—from *"a"* (without) and *"himsa"* (violence). In Book II of the *Yoga Sutras* of Patanjali, *ahimsa* is the first of the five "commandments" or *yama* (discussed in Sutras 30-31), which is further expanded in the five rules or *niyama* in Sutra 32.[4]

Over the years, the interpretation of *ahimsa* has expanded in terms of what is included. A number of present day religious groups share the value of avoiding violence—for example, the Amish, Jehovah's Witnesses, Mennonites, Quakers, and Seventh Day Adventists. And the Wiccan Rede states, "An ye harm none, do what ye will."

However, *ahimsa* has remained a concept defined by the absence of particular behaviors. We have shifted from a definition based on a lack of potency to a definition where the potency to act is assumed, but the focus is on what *not* to do. We still are not clear what positive action to take.

The Ethics of Reciprocity

Looking at a view of harmlessness that is more proactive, we find harmlessness equated with a code of conduct known as the ethics of reciprocity, or *treating others the way you would like to be treated*. This ethics of reciprocity is a common thread running through the tapestry of most of our spiritual traditions. Spiritual seekers around the world are admonished, one way or another, to "do to others what you would have them do to you." We see this in examples from a range of spiritual traditions.

When we stop to think about these ethical mandates, though, we realize that they are actually very self-focused. They leverage self-interest, starting from what we want for ourselves. Yes, they result in action rather than just an avoidance of action. But how do we know that what we want is actually in the best interests of the whole?

The philosopher and linguist Noam Chomsky has reframed this ethic slightly in suggesting that our most elementary moral principle is applying the same standard to all—or universality. "If an action is right (or wrong) for others, it is right (or wrong) for us. Those who do not rise to the minimal moral level of applying to themselves the standards they apply to others—more stringent ones, in fact—plainly cannot be taken seriously when they speak of . . . right and wrong, good and evil."⁵

Ethics of Reciprocity in Nine Spiritual Traditions

Bahá'í Faith	And if thine eyes be turned towards justice, choose thou for thy neighbor that which thou choosest for thyself. – Epistle to the Son of the Wolf
Buddhism	Hurt not others in ways that you yourself would find hurtful. – Udana-Varga 5:18
Christianity	So in everything, do to others what you would have them do to you, for this sums up the Law and the Prophets. – Matthew 7:12 (NIV)
Confucianism	Do not unto others that you would not have them do unto you. – Analects 15:23
Hinduism	This is the sum of duty: Do not do to others what would cause pain if done to you. – Mahabharata 5:1517
Islam	None of you [truly] believes until he wishes for his brother what he wishes for himself. – Number 13 of Imam "Al-Nawawi's Forty Hadiths"
Judaism	What is hateful to you, do not to your fellow man. This is the law; all the rest is commentary. – Talmud, Shabbat 31a
Taoism	Regard your neighbor's gain as your own gain, and your neighbor's loss as your own loss. – T'ai Shang Kan Ying P'ien
Zoroastrianism	Whatever is disagreeable to yourself do not do unto others. – Shayast-na-Shayast 13:29

The question with which quantum physics confronts us is what is "universal"? How far beyond our immediate human group does it extend? Does it include both genders? All nationalities? All living beings? Does it extend even further to include Gaia (our Earth) as a living entity and thus all of Gaia's component parts—mineral deposits and other components that we do not usually think of as "living?"

Positive Harmlessness

At the other end of the definitional spectrum from "innocuous," we find the term "positive harmlessness," which reflects a clearly proactive perspective. One of our maturational tasks is moving beyond self-interest and focusing on what is best for us all. This parallels a shift from a belief in our individual independence to an awareness of the Principle of Interdependence. Here the focus is on both what to avoid doing and what to do in order to ensure that neither oneself nor others is harmed.

Exercise: Behaving Harmlessly

Pick a recent situation in which you had the choice to behave either harmlessly or harmfully. How would you choose to behave if your aim was to avoid inflicting harm?

How would you choose to behave if your aim was to treat the other person the way you would like to be treated?

How would you choose to behave if your aim was to ensure the other was not harmed by yourself or others?

What difference, if any, is there in your three choices, and why?

In the metaphysical writings of Alice Bailey, we find harmlessness described as "harmlessness in speech and also in thought and consequently in action. It is a positive harmlessness, involving constant activity and watchfulness."[6] Further, "it concerns motive and involves the determination that the motive behind all activity is goodwill. That motive might lead to positive and sometimes disagreeable action or speech, but as harmlessness and goodwill condition the mental approach, nothing can eventuate but good."[7]

The Ageless Wisdom teachings just quoted from Alice Bailey are similar to the current day Native American Great Law of Peace: "Respect for all life is the foundation." Indeed, respect and an awareness of interdependence are crucial. We begin to see that positive harmlessness is more than just the ethics of reciprocity. It is motivated by goodwill or kindness, rather than self-interest, and involves firm action to remove obstacles and prevent harm.

Harmlessness and the Seven Principles

In order to understand what harmlessness is from a positive, dynamic perspective, it may help if we are clear about the nature of the cosmos in which we live and of which we are a part. In *Principles of Abundance for the Cosmic Citizen*,[8] we explored seven core principles that are central to our being and yet are often misunderstood. These principles are listed in the Preface and are directly relevant to our understanding of harmlessness.

First and foremost is the *Principle of Interconnectivity*, reminding us that we are all part of the same cosmic energy field. Most of us experience ourselves as unique individuals. In fact, establishing personal boundaries that separate and distinguish us from others has been assumed by psychologists to be a sign of maturity. But actually maturity is related to holding the tension of simultaneous separateness and wholeness.

Our ability to tolerate and even embrace harmfulness is rooted in the basic illusion that we are each separate individuals, unconnected to others except by emotional choice. This illusion allows us to believe that we can harm others without harming ourselves. We forget that, in a very real sense, there is no separate "self." As a Pima Indian proverb confirms, "Do not wrong or hate your neighbor. For it is not he who you wrong, but yourself."

But what does "harming others harms ourselves" mean in practical terms? Let's look at three examples:

> Think of a time when you felt mistreated or abused by someone else and wanted revenge. Recapture that vengeful feeling. Now look in the mirror — do you see a pleasant face or one where the eyes are strained and the face gaunt? Take your pulse or, even better, your blood pressure. Has it risen?
>
> Think of a time when you felt you definitely got the better of someone else (for example, cutting ahead in line, snatching the last item from under someone else's outstretched hand, grabbing the last parking space just before someone else tried to pull in). Recapture your sense of triumph and disdain for the other party. Now check your body — are you relaxed or tense? And try your blood pressure again.
>
> Think of a time when someone told you juicy gossip about someone you disliked. Recapture your sense of pleasure at that third person's humiliation, the sense that they deserved it. Again, try looking in the mirror

and checking for tension in your body. And try your blood pressure once more.

In each of these instances, the impact of our engagement in harmful activities is at least as great on ourselves as it was on our target. Much as we might like to believe that we are free to strike out at others without any effect to ourselves, it is not our reality.

> **Exercise: Interconnectivity and Harmlessness**
>
> Since we are all energetically interconnected, what does behaving harmlessly mean to you from a practical perspective?

We can help ourselves understand this dynamic by reflecting instead on how the energy of others affects ourselves. Can you remember a time when you encountered someone who really disliked you and who seemed to be "looking daggers" at you? Or can you remember how it felt when someone acted as though you were invisible and didn't exist? Conversely, can you remember how it feels to walk into a space where others love you and are eager for your presence? Even before a word is spoken or an action taken, you can feel that loving energy surrounding and enveloping you, can't you? These are just a few demonstrations of the energetic web that binds us together.

The *Principle of Participation* confirms that what we experience as reality is a result of our observation and the meaning we attribute to that observation.⁹ Our choices focus the energy on a particular outcome out of a range of possibilities.

An important aspect of our behavior is the perspective we bring due to the language we use. For example, how we name actions shapes our sense of harmfulness or harmlessness. Do you feel any different if you talk about "culling" elephant herds or "killing" elephants? If you refer to "missing women" or "abducted, violated women"? If you describe "procedures against Armenians" or "the genocide of Armenians"?

The linguist Benjamin Whorf argued that, since language presents reality in a culturally-specific way, we become parties to an unspoken agreement to organize our experience in those ways. All of us view our experience through the filter of a worldview — a framework of attitudes, beliefs, values, presuppositions through which we interpret our experience and interact with others. It is this worldview that shapes our perceptions, and it also provides us with answers to ethical questions about what is best to do in a given situation.

Exercise: Participation and Harmlessness

Name three ways in which your worldview affects what you consider to be harmless:

1.

2.

3.

On reflection, is there any aspect of your worldview you need to change if you are committed to positive (proactive) harmlessness?

To the concept of a worldview, the psychologist and linguist Susanne Cook-Greuter has added the idea of the "language habit which allows us to categorize and label . . .[objects and concepts]."[10] Cook-Greuter goes on to describe the language habit as having the following attributes:

- It constitutes a universal, all-pervasive dimension of human existence.
- It is innate but needs activation and modeling by expert speakers in early childhood to emerge.
- It is a learned behavior that becomes automatic and unconscious once acquired.
- It bundles the flux of sensory input and inner experience into labeled concepts shared with one's speech community.
- It is so deeply engrained that speakers of any given language are not aware of the reality construction imposed on them by their language.
- It can become a barrier to further [personal] development if it remains unconscious, automatic, and unexamined.

Recent research has shown that artists perceive the world differently than non-artists.[11] Artists tend to scan an entire environment, noting contours and colors and spending only part of their time on objects. Non-artists, by contrast, focus immediately on objects, turning images into concepts (of their own creation) through filtering and anticipation.

In an everyday context, what about when you are about to fix dinner? Until you decide what to fix, there are a number of possibilities. Once you decide, only one possibility becomes a reality.

Our perceptions and experience are directly shaped by our language and, in turn, create the reality that we experience. If we do not choose to become aware of the "language habit" or way of conceptualizing the world that we have acquired, we will have relatively little control over how we experience the world. In this instance, our "language habit" about whether or not to tolerate violence will shape our reality.

The *Principle of Nonlinearity* asserts that our experience, including how we change and grow, does not occur in a smooth sequential fashion. Time itself is elastic, and we have the ability to reframe our experience in ways that change and heal the past. We have already seen this with the Truth and Reconciliation process.

One of the critical aspects of this principle in relation to harmlessness is how we choose to name the cycles of our experience. Historically, we have marked cycles of change by the sequence of warfare (for example, after World War II) and the development of violence-related tools. How might our experience be different if we used time frames related to peace and harmlessness?

> **Exercise: Nonlinearity and Harmlessness**
>
> If you accept that the past and the future exist in the present—the "now"—how does that assumption affect how you would implement an ethic of harmlessness?

This brings us to the *Principle of Nonduality*, underscoring that our cosmos is much more complex than the simplicity that only two alternatives would imply. It also warns us about a key dynamic underlying violence, which is the division of the world into "us" and "them," or "friend" and "foe."

In order to harm someone, we have to be able to objectify them and experience them as "other" or "not self." Tales of genocide and mutilation are routinely accompanied by accounts of the ways in which the victims were first humiliated and degraded. As with the Principle of Interconnectivity, this principle reminds us of the interweaving of our energetic selves.

> **Exercise: Nonduality and Harmlessness**
>
> Pick a person you don't know (and likely won't meet) and visualize them as completely separate from yourself. Then visualize ignoring them or making a demeaning comment about them. How easy is that to do?
>
> Now visualize that stranger as energetically part of the cosmic whole of which you are also a part, seeing them as another version of the One Life. Then visualize ignoring them or making a demeaning comment about them. How easy is that to do?
>
> What impact does your sense of being "one," rather than "me" and "them," have on the behavior choices you make?

The *Principle of Interdependence* expands on the interconnectedness underlying our being. We are surrounded by a wide variety of life forms on whom we depend and who in turn depend on us. For example, we depend on plant life for oxygen just as it depends on us for carbon dioxide.

> **Exercise: Interdependence and Harmlessness**
>
> Walk outside in a natural environment (or visualize being there) and experience your interconnection with all the nonhuman living beings that surround you.
>
> If your wishes were to appear to run counter to those of other living beings (for example, wanting to pick a flower rather than leaving it rooted in the ground), how could you handle the situation in a harmless, rather than harmful, manner?

In our cosmos, all other beings matter and deserve respect. We are reminded to refrain from harming all other life forms—not simply other humans—through thoughtless or deliberate dismissal and to thank them for their contributions.

With the *Principle of Adaptability*, we see that we are part of an ongoing process of growth and change. There is no necessary failure, only feedback and learning what not to do the next time.

We grow through the process of focusing and relaxing tension. The question becomes what range of choices we permit ourselves for that release. Do we "let off steam" by allowing ourselves to be violent? Or do we take responsibility for managing our own energy field in a harmless manner?

> **Exercise: Adaptability and Harmlessness**
>
> Picture being very tired and irritable and then being interrupted in a task that is time-sensitive and requires your careful attention.
>
> How might you handle your frustration at the interruption in a manner that is harmless rather than harmful?

Finally, the *Principle of Cooperation* directly challenges the widely held belief that we are innately violent, mired in self-interest, and that competition is at the root of our survival. Scientists like Lynn Margulis have shown us that networking, not competition, is actually the fundamental survival strategy used from microbes on up.[12]

> **Exercise: Cooperation and Harmlessness**
>
> Picture yourself enrolled in a competition for a contract that you very much want to win. There are four other highly qualified competitors.
>
> What options do you have for approaching the situation in a harmless manner?

Recent psychological research has verified that other primates are helpful, cooperative, and sensitive to situations where participants are not treated equitably.[13] Of course, we know from watching children that, if we expect someone to behave violently, they will likely do so. If, on the other hand, we trust that others will cooperate, our trust will often be rewarded.

Harmlessness in Thought

We know, from the Principle of Interconnectivity, that our thoughts are part of our energetic entanglement. We actually have as much power to harm ourselves and others through what we think as through our visible actions.

The basic building blocks of our thoughts—which determine what we say and do—are our values, attitudes, and beliefs. We have a responsibility to value the control of impulses, to observe or monitor our behavior, to weigh options and consequences before we act, and—perhaps most importantly—to screen out messages that tell us we are entitled to act harmfully, both to others and to ourselves.

Harmlessness in thought also involves the matter of intention. This is, of necessity, complex. The 1981 movie *Absence of Malice*, with Paul Newman and Sally Field, underscored the damage that can be done to innocent people by someone—in this case a reporter—with the best of intentions but inaccurate data.

Quantum physics has now demonstrated that the energy field in which we exist is multi-potential—like a mist of possibilities—and that it is our observation, or intention, that literally precipitates the final selection and thus creates reality.[14] A way to help visualize this is to think about the process of helping someone (including yourself) apply to a number of colleges. Until a decision is made, all of the possibilities exist; but, as soon as a selection is made, one actuality emerges and the other possibilities fade away. In this way, we continuously have the ability to choose positive options in support of joyous growth.

Exercise: Harmlessness in Thought

Give three recent examples of when you have been purposely harmless in thought:

1.

2.

3.

How did you feel as a result?

Harmlessness in thought involves the type of focus we choose to maintain. It is only in the past ten years that there has been research on the consequences of a positive, rather than a negative, focus. The positive psychology initiative was launched by Martin Seligman when he became president of the American Psychlogical Association. There is now a Positive Psychology Center at the University of Pennsylvania, a *Character Strengths and Virtues Handbook,* and an International Positive Psychology Association. What is particularly relevant to us is the research that is showing that happiness and a positive attitude result in an openness to new experiences and increased creativity. We can contribute to a creative, healthy environment by disciplining ourselves to maintain a positive focus.

Harmlessness in Word

There is an old saying, "Sticks and stones may break my bones, but names will never hurt me." That might true if we did not internalize the names that are intended to be hurtful. But, in fact, sound and the spoken word are extremely powerful. In the Christian Bible, the book of John begins with, "In the beginning was the Word" — the sound that initiates. We can witness the power of sound when a highly-trained soprano is able to shatter a wine glass with a single high note.

Exercise: Harmlessness in Word

Give three recent examples of when you have been purposely harmless in what you said or did not say:

1.

2.

3.

How did you feel as a result?

The language that we use defines how others see us. In the 1960s and 1970s, there was a great deal of research conducted by anthropologists and psychologists regarding the impact of language on how we view the world. As a result, we have become aware that inclusiveness itself needs to be grounded in inclusive or bias-free language. We have already discussed the "language habit" and its role in shaping our experience. One of the ways that we disrupt negative, harmful effects of language is by renaming. Thus, we see the shift from the derogatory "colored" to the empowered "Black," from the demeaning "girl" (when referring to an adult) to the more respectful "woman."

A second example of the importance of the words we select is illustrated in a simple exercise. See if you feel the same if you say, "I'd like to join your committee, and I have some ideas about how to make it more inclusive," or if you say, "I'd like to join your committee, but it needs to be more inclusive." Generally speaking, when we use "but," we imply controversy and an adversarial position. We tend to become tense and to focus on how to justify our position or approach. When we use "and," we imply an openness to alternatives and become more cooperative.

Finally, while words are important, so is silence—from two perspectives. There are times when it is *important* to remain silent. We have no right, for example, to offer unsolicited criticism to others since we don't know the life or Soul purpose of any other person or what their life journey has been to date. Or, if we have no actual data on a controversial issue, we have no business chiming in to agree just so that others will like and accept us. Or, if by speaking out we would undermine the authority of the person who is in charge of a particular initiative and is behaving without harm, then we need to keep silent.

And then there are situations where remaining silent condones harmful actions and it is important to speak out—for example, in protesting unjust wars or ecological degradation. The same is true when another person is being unjustly maligned or negatively stereotyped—if we remain silent we are in essence agreeing.

Speaking out can have positive results, especially these days when social media are used. The Wake Up Call initiative on September 21, 2009 regarding the need for an effective climate treaty, for example, resulted in 2,632 events in 134 countries and certainly captured the attention of political leaders.

Harmlessness in Action

One of the key components of harmlessness in action is the ability to set limits and to eliminate outmoded structures and practices. This is tricky. We need to be careful not to equate harmlessness with being kind and friendly. Sometimes, as with speech, we need to act to prevent harm from occurring. As parents, we know the importance of setting limits so that children are not harmed, though the children may be angry about those limits. Or, as the Dalai Lama has indicated—in response to a question about whether killing another was ever justified—sometimes we need to act (with compassion) in order to keep others from generating more negative karma.[15]

Similarly, we know that a critical component of the ability to create is the ability to destroy or wipe away old forms. In the metaphysical writings, the cosmic energetic field is described as having seven main attributes or characteristics, known as the Rays. Ray One—the Ray of Power—is said to combine both creative and destructive aspects, illustrating that beginnings and endings go hand in hand and underscoring that ending is not necessarily harmful.[16]

Another component of harmlessness in action is the process that we use. The new forms demand an inclusive, consultative process. We see this happening already in the way that so many ordinary people are engaged in social media such as Twitter and Facebook. During the last presidential campaign in the United States, Barack Obama's use of information technology and social media was an excellent example of how to include a wide range of voters. One of the ways that we avoid being harmful is by actively engaging others in determining what is in their best interest, instead of deciding for them. We want to be sure to be working towards a positive goal rather than away from a negative one. Deepak Chopra's "Peace Is the Way" Global Community demonstrates the positive process of consciously working towards peace rather than against war.

> **Exercise: Harmlessness in Action**
>
> Give three recent examples of when you have been purposely harmless in how you acted:
>
> 1.
>
> 2.
>
> 3.
>
> How did you feel as a result?

This brings us to what we could create if we focused on harmlessness. Here are some ideas about common institutional structures:

Education:

> In British Columbia, Canada, experiments are moving forward regarding the inclusion of social responsibility and mindfulness in the school curriculum, starting at an early age. This is part of a focus on heart mind education and strengthening mindful attention.[17] It could be part of a training program in positive harmlessness.

Health:

> We are already seeing a shift in focus from disease to health, from treatment to prevention. We see energy work being validated along with more traditional approaches. Information technology now allows us as patients to become more actively engaged and in charge of our own medical records. As we are more informed, we are in a position to reduce harm from "iatrogenic disease," or disease caused by misdiagnosis or inappropriate treatment by doctors.

Religion:

> The United Religions Initiative provides a model for working together across different systems of spiritual belief. Using the method of appreciative inquiry, it stimulates curiosity about persons from different belief systems. The approach of appreciative inquiry is becoming so widespread that there was a conference in Nepal in November 2009 called "World Appreciative Inquiry Conference: Creating a Positive Revolution for Sustainable Change."

Media:

> All too often the negative is reported without also reporting the positive. In fact, some journalists reject positive news, saying that it is free advertising! There are fortunately already some leaders in ensuring that the

positive is reported: Good News Agency, *Yes*, the *Utne Reader*. Hearing the positive is a critical component of underscoring the value of harmlessness.

Government:

There is already an international initiative to establish Departments of Peace. The challenge will be to have them replace Departments of Defense or Homeland Security rather than to exist in parallel. Another idea is to replace Departments of Foreign Affairs or Immigration with Departments of Multinational Cooperation. And there are already proposals that governments could be operated with the participation of all citizens, via the Internet, instead of needing to elect representatives.

Valuing Positive Harmlessness

We are indeed on the threshold of a global shift in consciousness. How will we make choices about what we do or don't do, how we do it, and why? We have the opportunity to shift from rhetoric about the importance of harmlessness to demonstrating positive harmlessness as the essential defining characteristic of our humanity.

As we create the organizations, groups, and instruments that express the new ideas and values, the principle of positive harmlessness needs to be central to our process. We need to value harmlessness as the attitude of greatest strength, requiring self-discipline and fortitude.

THREE

Harmlessness and the Butterfly Shift

We value harmlessness. Yet that valuing, by itself, is not enough for us to put harmlessness into consistent practice. Behaving harmfully permeates so much of our lives that we take some level of ongoing harm for granted. We almost don't see it!

When we reflect on the escalating violence or harmfulness in our world, we know that significant changes are required. Damage to our environment through greed and entitlement is jeopardizing our home — the biosphere that is our Earth, Gaia. Damage to our societies through armed conflict is swelling the ranks of refugees and violated persons. Damage to the socio-economic fabric within which we live interdependently is widening the gulf between the middle class and the very poor. And that is only the tip of the damage to living beings throughout our world. How is it that we are still allowing ongoing "gendercide" without international sanctions, which is responsible for the murder of at least 100 million baby girls every year?[1]

What will it take for us to begin to truly practice positive harmlessness when we are so used to an environment of harm? How can we become accustomed to positive harmlessness, rather than harm, as our norm? What will help us move towards an internally-defined ethic of harmlessness that can shift us, individually and collectively, away from violence?

Simply deciding that a change is needed is not enough. We all know this from making New Year's resolutions that don't last. For example, we may be concerned about being out of shape or overweight and so we resolve to exercise daily, eat only healthy foods, and drink lots of water. Typically, we are back to our old patterns within a month.

The problem is that lasting change is not generated from fear. Fear simply makes us want to avoid thinking about what's happened. Lasting change springs from choice—and from making that same choice over and over again.

In order to change, we must first be aware of a need for change. Often that need is triggered by a personal crisis. We reach a point where we say, in some fashion, "I can't go on this way." When a number of people share that same sense of crisis, we reach a tipping point where broader social change is possible.

The kind of change necessary to shift from harm to harmlessness as our ethic is particularly challenging. We have no shared experience of living harmlessly. It is difficult to choose a way of being that we have not seen modeled. We need to create for ourselves a collective experience of harmlessness, a shared expectation that we will automatically and reflexively behave in a harmless manner.

To begin, we will be examining how we usually initiate a change process and what is required for that change process to become engrained as habit. Then we will introduce a daily practice that we can use to begin to create a familiarity with the feeling of harmlessness.

Becoming Conscious of the Need to Change

As long as we are comfortable with things as they are, we will not seek out new options. So how do we become uncomfortable with the familiar, the status quo? How do we become aware that our attitude or behavior is inappropriate, just an old habit that no longer serves us well? How does that initial question arise? How do our values suddenly get challenged?

Generally speaking, we change because our old belief structure no longer fits who we are, because we have experiences that confront us with gaps between how we think of ourselves and how we are actually behaving. The 1967 classic movie *Guess Who's Coming to Dinner?* is an excellent illustration of precisely this dynamic. Two Caucasian "liberals" (Spencer Tracy and Katharine Hepburn) are confronted by their daughter with a potential son-in-law (Sidney Poitier) who is Black and discover that they are more racist than they realized. If they want to actually live their stated belief that ethnicity doesn't matter, they will need to change their attitudes and behavior. Our challenge is to create a similar dissonance between harm*ful*ness and harm*less*ness.

Exercise: Recognizing the Need for Change

Pick a value that is important to you.

Now, keeping that value in mind, review the past 48 hours carefully. Did you always act in keeping with that value?

If you did, what helped you do so?

If you didn't, when didn't you and why?

In order to make a change, we need to be conscious of the beliefs, attitudes, or behaviors that are involved. This is not as easy as it sounds. Some of our limiting beliefs come easily into our awareness. But there is another deeper level outside of our conscious awareness, the subconscious. Here we find limiting beliefs that were formed when we were very young, often before we had language. While we usually assume that language is essential for attitude formation, we have only to reflect on the vast number of learning tasks that preverbal infants master to realize that learning is not language dependent. One of the fascinating recent studies of newborns demonstrated that those whose mothers were bilingual and spoke both languages regularly during pregnancy showed a strong interest in, and ability to discriminate between, the two languages shortly after birth.[2]

Some of our beliefs or behaviors are unquestioned because they are appropriate in one setting or under one set of circumstances even though they aren't in others. For example, suppose you are known as a good, safe driver. What does that mean? If you drive an automatic transmission car in North America, would you necessarily be a safe driver in Great Britain where they drive on the left rather than the right? Would you be able to drive smoothly in a standard transmission car or a truck? Sometimes our assumption that all is well comes from automatically and inaccurately generalizing from a very specific circumstance.

We also have limiting beliefs, most commonly expressed as stereotypes, that are formed as a way of coping with the daily onslaught of data. It is impossible and even counterproductive for us to deal with each bit of data as unique. For example, we would have difficulty making sense of letter sequences like, "I a m g o i n g t o b o r r o w t h e c a r" or "P l e a s e s i t d o w n" on a continuous basis. Treating each letter separately eliminates the overall sense of meaning. Once we group the letters, though, we get, "I am going to borrow the car" and "Please sit down." Both of these sentences make sense. The sense or meaning comes from the process known as "tokenization" or forming clusters of data. Seen from this perspective, stereotypes serve a useful function. They provide us with a shorthand for how we are likely to experience a new situation.

But stereotypes can become counterproductive through oversimplification. Take, as an example of generalization and data compression, the matter of mathematical rounding. By tradition, we would round both 21.76 and 21.84 to 21.8. We would also round 21.8 and 22.4 to 22.0. But are 21.76 (our original number) and 22.4 really the same?

Stereotyping helps us, but at a cost. Stereotypes assume that similar situations or people are identical when often they are not. The stereotype typically contains a judgment, either positive ("all Asians are good at math") or negative ("women aren't as strong as men"), that makes our perceptions less flexible. So we need to be aware of stereotypes and question their applicability in specific situations—which isn't always easy.

Research has shown that stereotypes can form without any actual experience with the stereotyped group and that children have internalized negative ethnic stereotypes as early as five years old.[3] As well, gender stereotypes regarding math and science are still widely held around the world.[4]

So, by the time we are adults, we already have a wide range of unconscious limiting beliefs and stereotypes that hold familiar patterns in place. Unfortunately, those familiar patterns—those "givens" that we take for granted—assume a context of harm or violence, whether towards others or towards ourselves.

Awareness of the need for change is the first step towards making that choice. It is also the most difficult. It means consciously choosing to move out of our comfort zone of habits and reflexive unconscious actions. It means resisting all our strategies to maintain the status quo. It means committing to, and valuing, an ongoing change and growth process.

There are at least three ways that we become aware of a need to question our current behavior. *First*, we might observe a role model, someone we respect, behaving unexpectedly. For example, that role model might behave humbly in a situation where others demand special treatment. Or the role model might refuse to laugh at a racist or sexist joke, enduring the awkwardness of not acting like "one of the crowd." Suddenly we become aware that there are other options, ones that might be more in line with our own values.

Second, we might become aware through contrasts between what we value and what is happening around us. The greater the contrast, the more likely we are to question our assumptions. For example, if we were to place a white dot on a white background, it is virtually impossible for us to see it. It is not differentiated enough from its surroundings. If, however, we place a white dot against a black or a blue or a multicolored background, the dot shows up very nicely.

The role of contrasts is illustrated in the folklore about frogs and boiling water. It is said (not necessarily correctly) that if you put a frog in boiling water, it will hop right out and thus save itself. But if you put it in cold water that is gradually heated, it won't notice the difference until too late and will boil to death. Heightened contrast helps focus our attention and stops us from continuing to act mindlessly, so to speak. This is why sometimes a social injustice or a violent dynamic can go unnoticed until suddenly it becomes so blatant that we can no longer ignore it.

> **Exercise: Identifying Cues for Change**
>
> Think back to a permanent change that you made in your attitudes and actions. What cue or cues precipitated your questioning your old attitudes and actions?

Third, we might use a regular practice of self-reflection to examine our actions against our values. For example, we may engage in a regular evening review as part of a spiritual practice. Or we may have the self-discipline to stop when we are feeling tense or hopeless and ask what personal incongruity is responsible for that feeling. Or we might set ourselves guidelines for self-examination such as asking ourselves, whenever we start thinking in an inappropriately linear or dualistic manner, what might be some other options.

Our Incentive to Change

Becoming aware of the potential for change is not enough. We need an incentive to actually make a change. Psychological theory tells us that we change when the perceived costs of remaining the same become too high. Usually there is a precipitating crisis for this reactive strategy. Traditional economic theory tells us that we change out of self-interest in order to increase profits. But recent studies bridging psychology and economics show that people have a natural inclination to be cooperative and that people tend to value equity (or fairness) over economic efficiency (or higher profit).[5]

To understand these new findings, let us examine the behavior of children. Children kept confined in cribs without human contact—as has been the case with Romanian orphans—emerge lethargic and psychologically impaired.[6] By contrast, children raised in a loving environment are active and inquisitive, continually trying new things and testing how things work. They are attracted by the potential for growth.

Generally speaking, we don't respond well to being told that we have to change. What works better is to become engaged in a process (or game) of experimentation so that we begin to experience firsthand the benefits of the new approach.

> **Exercise: Motivation to Change**
>
> Think of the last two personal changes you made. What was your incentive?
>
> Is this an incentive that works for you in a range of situations?
>
> If so, how can you strengthen it?
>
> If not, what other incentives could work better for you?

Of course, sometimes it is not possible to engage ourselves or others in a needed shift because we feel too vested in the status quo. In that case, the incentive to change will need to be provided externally. One option is that those around us who wish us to change no longer tolerate our habitual behavior or rationales — the "tough love" approach. Another option is that we are placed in situations repeatedly where our habitual behavior is in the minority — the "peer pressure" approach.

Or, at the extreme, legislation is passed that requires us to change.

Changing long-held patterns of behavior is challenging and takes effort so we need to have a sense of choice and of support. One option that can work well is to act "as if" for a period of time in order to try out different options and see if they will work for us. As adults, we are usually entranced by the opportunity to take free quizzes or scales, to work puzzles, or to play games. We enjoy testing ourselves against unknown circumstances or obstacles for the pure pleasure of stretching our capacities and learning something new. This thirst for discovery can become an incentive to change ingrained behaviors.

Making a Change Permanent

Change is a familiar and ongoing process. We see change modeled for us in the cycles of nature around us. Our lives literally depend on change as tens of thousands of cells in our bodies are replaced each day. We also see the negative effects of no change in dynamics like stagnant water that becomes unhealthy for us. However, we also resist change because we want a sense of stability and continuity.

In sports, we have numerous examples of how specific occurrences have shifted expectations permanently—for example, breaking the four-minute mile record for running, or executing the "quad" in figure skating. Generally speaking, the changes in social behavior that have lasted are those that are enforced by law—for example, the wearing of seat belts or not smoking in designated areas. But if the law were removed, would our behavior revert?

In many instances, what appears initially to be a fundamental change gradually shifts back—like a forest reclaiming land after it has been cleared, built on, and then abandoned. Or the shift back can be consciously stimulated by advertising messages. For example, when men enlisted in the military during World War II, women performed virtually every economic role in western civilian society. But, when the war ended, advertising and other media began emphasizing the importance of women remaining in the home (and leaving paid jobs to returning veterans); and workplace gender prejudice re-emerged.

One of the simplest revelatory activities about how change can revert is our choice of language. Prior to the 1960s, we may have been able to plead ignorance. But extensive research in the 1960s and 1970s confirmed the work of Whorf, showing that the language we use shapes how we see the world. After persistent pressure to change from gender-specific to gender-neutral language, we now no longer have the want ads divided by gender. We now refer easily to "flight attendants" and "letter carriers."

But we are gradually reverting to the use of "girls" or "ladies" to refer to mature women in contexts where we would never refer to males in similar roles as "boys" or "gentlemen." In the 2010 Winter Olympics, for example, many of the women's events were labeled "ladies" (Men's Snowboard vs. Ladies' Snowboard, Men's Moguls vs. Ladies' Moguls, Men's Speedskating vs. Ladies' Speedskating, and so on), and the announcers repeatedly referred to the "girls" competing. Many of the gender-neutral options no longer come readily to the tongue. For example, instead of substituting "staff the booth" for "man the booth," people either allow the masculine phrase to be used unchallenged or struggle laughingly with awkward substitutions like "woman the booth."

So what creates lasting change? We need an experience of a new way of being that is so attractive and feels so right that we are willing to give up familiar habits and engrain new patterns. Recognizing the need to experience a new way of being in order to sustain change lies at the heart of immersion experiences. We see this dynamic played out in workshops that are hailed as life-changing, where participants experience themselves in a new context. The more comprehensive the experience, the more likely our limiting beliefs and stereotypes will shift to allow for new possibilities.

> **Exercise: Change Process**
>
> Choose an important and difficult change that you made in the last year and reflect on the following questions:
>
> 1. What was the problem that you were trying to resolve by making the change?
>
> 2. What was your motivation? Why did you consider making a change?
>
> 3. Why were you able to make that change? What were the components of your success?
>
> 4. Did you make the change all at once, or did you "backslide" several times to the old behavior?

Some examples may help us understand what we need to create. Language immersion programs, especially those with a host family home-stay component, increase successful learning of a new language because they shift participants out of their normal linguistic environment and provide a context where a different language and culture are the norm. Olivia Travel helps travelers and cruise staff alike become less homophobic by providing all-lesbian cruises where being a lesbian is the norm and lesbian couples are treated publicly as couples. Substance addiction treatment programs are residential on purpose in order to create a sustained experience of life without that substance.

So, in the absence of "harmlessness immersion" programs, we need to find ways to create that sense of "ah ha, so this is what it is like." Such experiences are essential to our being able to visualize and embrace the changes necessary for us to establish harmlessness as a habit.

Introducing the Butterfly Shift Dynamic

For us as individuals, shifting humanity's focus from harm to harmlessness is overwhelming. Just imagine stopping warfare, eliminating violence against women, ending poverty, ensuring access to clean water, expanding biodiversity, reversing global warming—is it even possible? In the face of such incredible odds, it is not surprising that many of us take few of the steps available to stop violence and its companions—greed and a sense of entitlement to over-consume.

How then are we to bring about this lasting change? To begin with, we need some parallel to a harmlessness immersion process that we can begin bit by bit. We need a simple daily practice that will (a) disrupt our familiar but harmful patterns, (b) engage us immediately with a sense of hopefulness, and (c) have a high probability of leading to new harmless behavior. We need a parallel to learning to fish rather than being given a fish. If we can begin to experience what it is like to have harmlessness as our primary experience, we will be in a position to move towards harmlessness as our ethic. Enter the concept of the butterfly effect.

Mathematician and meteorologist Edward Lorenz popularized the term "butterfly effect" in his 1972 paper for the American Association for the Advancement of Science titled,

"Predictability: Does the Flap of a Butterfly's Wings in Brazil Set Off a Tornado in Texas?" For our purposes, there are several implications of this chaos theory dynamic. *First*, there is the general concept of a ripple effect spreading out from a small, seemingly insignificant action. We see this when we drop a pebble in a pond and watch the ripples widen from that initial contained splash. *Second*, there is the notion of sensitive dependence on initial conditions. The smallest change in that initial small action produces large variations in the end result. So we see the ripple pattern change if we throw the pebble out to the middle of the pond instead of dropping it close to shore. Similarly, the flap of the butterfly's wings creates a minute shift in atmosphere such that a large-scale event like a tornado could be precipitated or averted. *Third*, the relationship between the initial action and the dramatic end result is not necessarily causal. The butterfly doesn't flap its wings in order to create or prevent a tornado. The initial action simply sets the stage for possibilities.

The notion of ending violence and shifting the course of human history through leveraging the butterfly effect—a "Butterfly Shift," if you will—is both immensely appealing and potentially possible. It depends on the ramifications of the Principle of Nonlinearity, from which we can deduce that, no matter the scientific data, it is never too late to initiate change. Our options may become modified or more limited the longer a situation remains unresolved, but we can always choose to shift the energetic context holding it in place.

There are many initiatives that are already helping to shift our energetic environment in a positive direction—for example, the practice of random acts of kindness, affirmations, and meditation itself. All too often, though, we are giving others positive energy without teaching them to participate in generating it. And so the end result is unnecessarily limited.

The Butterfly Shift is not intended as a quick fix or a panacea for all our woes. Our challenges are far too complex for a single solution. Rather, we can think of the Butterfly Shift as giving us experience in gradually substituting positive, harmless energy for current harmfulness. Think of a pail of dirty water (the harmfulness). Now imagine clear, clean water (our Butterfly Shift activity) pouring in and displacing or crowding out that dirty water until the pail is full of clean water. That is the consequence being proposed.

So what are the prerequisites for the Butterfly Shift dynamic to provide us with a useful mini-immersion in harmlessness? Early work on attitudinal change[7] taught us that attitudes themselves have a cognitive (or belief) component, an affective (or emotional) component, and a behavioral (or action) component. The most critical of these is the emotional component because it is the evaluative dimension and carries with it a charge of deeply-held feelings. It is what drives us to say things like, "I don't know why but I'm just not comfortable doing that" or "I know it doesn't make sense but I just hate them" or "I just have to have it no matter what."

Gandhi once said, "Be the change you want to see in the world." We can learn to be that change by using a daily Butterfly Shift practice to create familiarity with a milieu of harmlessness. This will reinforce a sense of relatedness to, and respect for, others while instilling the beliefs and attitudes that underlie a sense of abundance and crowd out violence. We cannot force others to change because they have free choice, but we can initiate the energy that makes change possible and even likely.

The potential for success of the Butterfly Shift as a mini-immersion experience has been verified in the recent research on "joint attention" or the fact that people spontaneously follow the gaze or focus of other people.[8] Even as infants, we learn new information or ways of being by observing what others do. In fact, on a neurological level, observing a role model has the same effect as doing the behavior yourself![9]

Three additional contributions that the Butterfly Shift dynamic offers are: (a) a conscious focus on the helping relationships that we often take for granted, (b) a method for expanding our ability to contribute positive rather than negative emotional energy, and (c) specific positive feedback to others about how they have transformed our lives through their helpfulness in order to help formulate an ethic of harmlessness. This last step is particularly important because that feedback is what makes it possible for others to replicate their empowering actions if they so choose.

The most powerful dynamic we know is that of feeling seen and heard. Economic analyst David Korten has noted the power of conversation and of being heard in an emotionally supportive environment.[10] The feminist theologian Nelle Morton, in *The Journey Is Home*, referred to this as hearing each other into existence.

The Butterfly Shift

While we retain responsibility for harmlessness in thought, word, and action, we need to do more if we are to build an alternative to our current violent culture. As long as violence is ubiquitous, it is hard to imagine a different reality. We can begin to create a "harmlessness mini-immersion" experience by translating the power of feeling seen and heard into a daily three-step Butterfly Shift practice:

Step One: Notice the Shift Potential

Be clear about the situation in which we wish to take action as well as the action possibilities.

Example: Notice the name of the service person being more helpful than required.

Step Two: Feel the Shift Potential

Recognize the possibilities for imbuing our action with a strong positive emotional energy, making it more potent.

Example: Feel grateful for the service person's help.

Step Three: Act on the Shift Potential

Take a positive action that is personal and specific.

Example: Thank the service person by name for the help, specifying how they were helpful to you and what difference it made.

Each of these three steps will be described in detail in the next chapters.

Notice that the examples given above all provide opportunities to express respect and gratitude to a service person we might otherwise disregard and who has been particularly helpful to us. This is only one type of Butterfly Shift encounter—what we might call the "grateful" type. We can identify three types of Shifts altogether:

1. The *Compassionate Shift*

    Here our focus is on supporting others in shifting from being negative, due to exhaustion or distress, to being able to be pleasant while doing their jobs.

2. The *Grateful Shift*

    Here our focus is on helping others to replicate the excellent assistance they have provided.

3. The *Joyous Shift*

    Here our focus is on encouraging others to pass on the positive feeling we have expressed toward them.

As we examine the three steps of the Butterfly Shift in the next chapters, we will be noting what is similar and what is different for each of the types of Shifts.

Leveraging the Butterfly Shift

Because we are aware of how intimately harm is woven into our lives, we can see that the Butterfly Shift is an essential step in reshaping our expectations. It provides us with a context in which we can consciously experience harmlessness as the keynote of our lives, individually and collectively.

Once we are clear about how to simulate a harmlessness immersion through the Butterfly Shift dynamic, we will move on to explore what this new perspective suggests as an alternative model of adult maturation. We will outline how to ensure that harmlessness is a natural outgrowth of personal development, minimizing the need to recover from socialization into a norm of harmfulness and establishing a shared ethic of positive harmlessness.

Lasting change occurs on a personal level when we become so used to a new way of being that we lose interest in old patterns. By engaging in the daily Butterfly Shift, we begin to change what we believe is possible and appropriate. We release our addiction to violence and become accustomed to positive harmlessness as our usual way of being. And we create the opportunity for others to change as well.

Part Two

Immersion in Harmlessness—The Butterfly Shift

FOUR

Managing Our Focus

We choose what we notice. Every minute we are bombarded by hundreds of stimuli competing for our attention. To have an effect and create a change, we first have to actually pay attention to what is happening.

It may seem obvious, but what we notice is not determined automatically. What about you as you read this page? Clearly you are noticing the words on the page. But what in addition to those words? Are you noticing the font — its size and readability — or the page layout and line spacing? Perhaps, in addition to the words, you are aware of your body and how comfortable you are, where there is pressure, what your body temperature is. Or perhaps you are aware of your surroundings — voices or music in the background, visual cues in the room or out a window glimpsed from the corner of your eye.

This first step in the Butterfly Shift mini-immersion dynamic of *noticing* requires us to select a situation where we can take action and decide what the action possibilities might be. The situation needs to be one where we have an emotional response — particularly compassion or gratitude or joy — and are directly involved.

We talk about choice as though it were a discrete process that occurs occasionally. In actuality, we are making choices all the time. On a perceptual level, we are continually selecting what we will look at or listen to. If we are reliable, we choose to be on time to work each day. So we are actually experts on making choices. Even if we make no decision, we are choosing not to choose.

But some of our choice options may be beyond our awareness. On a physical level, for example, we choose unconsciously to keep breathing, to keep our heart pumping, to digest our food, to blink and keep our eyes moistened, and hundreds of other critical bodily functions.

While we may have practice in choosing, we might miss the small action opportunities that are critical to success with the Butterfly Shift dynamic. To notice means to perceive or to observe. But simple perception is not enough. We shall see that being engaged, not just passively perceiving, is necessary if we want the Butterfly Shift dynamic to work.

Focusing Our Attention

The Principle of Participation reminds us that our perceptions create our experience of reality. Perception involves concentrating or paying attention. The psychologist William James is credited with the still-current definition of attention: "Everyone knows what attention is. It is the taking possession by the mind, in clear and vivid form, of one out of what seem several simultaneously possible objects or trains of thought. . . It implies withdrawal from some things in order to deal effectively with others."[1]

Psychologist George Miller has shown that we can only attend consciously to five to nine items simultaneously, with the magic average being seven.[2] Everything else we perceive is processed by our unconscious mind, including a wide range of nonverbal cues.

The screening-out process that we use, as we focus on those seven critical items, is in part a narrowing of attention. It is a purposeful (though not necessarily conscious) limiting of the number of cues that we are noticing.[3] This narrowing happens particularly when we are emotionally aroused—angry, jubilant—and appears to be related to the narrowing of attention that accompanies the adrenaline rush from fear. It urges us to zero in on the most critical aspects of the situation.

But recent research suggests that our perceptual field may become even narrower than we've thought.[4] If we don't give something our specific attention, it is as though it just didn't exist. And, in the sense underlying the Principle of Participation, it actually doesn't! Our limited capacity for focused attention, which selects the amount of information we process at a given time, makes us disregard other details. We have terms for this, like "in her own world" or "lost in my book." So we may screen out and not notice critical, relevant behaviors.

Exercise: Absorption

Pick a recent activity when you were completely absorbed.

Why were you so absorbed?

What did, or would, make it possible for you to shift out of that absorption?

When we screen out other things that are happening, this "overlooking" may be benign—that is, we may be operating as though on automatic pilot, becoming alert only if something out of the ordinary occurs. Have you ever had the experience of driving home from work or from the grocery store—along any familiar route—and having no memory of the drive when you arrived home safely?

"Change blindness" is a term that was introduced in 1997 to explain a less benign process of missing large changes in our visual field. This can happen when there is a visual disruption just at the moment of change.[5] So we miss details to which we are not paying direct attention. As an example, have you ever read a manuscript or book where a word was missing at the turn of the page and found that you automatically filled in the word? This is an example of change blindness.

> **Exercise: Shifting Attention**
>
> Pick a time when someone you know well is so absorbed in a task that they don't notice you. How can you get their attention?
>
> When you do get their attention, how do they react (pleased, irritated, curious)?
>
> Why do you think they have that emotional response?

Psychologists Arien Mack and Irvin Rock coined the phrase "inattentional blindness," which is a potentially more serious outcome of preoccupation.[6] They found that, if people were sufficiently absorbed in what they were looking at, they completely missed other very noticeable, and even quite bizarre, visual cues. So, if we are not careful, we can literally miss seeing things that are right in front of us. Also known as "perceptual blindness," inattentional blindness occurs either because we are distracted by something else or because we have no frame of reference for what we are seeing (and so it seems irrelevant to us).

The advent of cell phones and their pervasiveness has renewed research interest in this dynamic, which is now being called "cognitive capture" or "cognitive tunneling." In 2003, research began to relate cognitive capture to the activity of driving a car (during a simulation) while talking on a cell phone. Participants in the simulation had very limited recall of road signs that they passed and even rear-ended cars braking in front of them.[7] Follow-up research in 2006 found that talking on a cell phone while driving actually impaired a person's driving skills more than being drunk.[8]

The vivid world of the iPod has also been implicated in cognitive capture, with persons being killed as they stepped out in front of traffic, oblivious to the danger because of their absorption in the multi-media world of the iPod.[9] The number of accidents due to following GPS (global positioning system) data rather than looking at the road has been increasing, with the world inside the Blackberry becoming more vivid and engaging every day — to the detriment of highway safety.

Our inattention to our immediate surroundings might be due to any one of a number of sources, not only cell phone or iPod involvement. We may be internally focused because of some issue we are trying to resolve — a fight we just had, how to meet a looming deadline, or how to develop a concept for a lecture. We may be caught up in talking with a friend or family member. Or we may be on sensory overload and trying to eliminate unnecessary details.

So the message to us is that noticing requires specific focused attention. Indeed, we could say that there is no conscious perception without our choosing to pay attention.

Exercise: Focused Attention

Place yourself on a bench or seat outside where you can safely ignore what is going on around you for a few moments. Focus internally on your most recent conversation with a good friend, remembering the nuances of how the person looked and seemed to feel.

Now shift and become aware of your surroundings in as much detail as possible. Notice sounds made by insects or small animals or the wind. Notice smells and visual details.

When you are focused on your surroundings, how easy is it to remember the details of being with your friend?

When you were engaged with the memory of your friend, how much did you actually notice of your surroundings?

What does this exercise illustrate for you about the focus of your attention?

Mindfulness and Mindlessness

Our own process influences what we notice. Social psychologist Ellen Langer has focused her research on the difference between mindlessness and mindfulness.[10] Mindlessness refers to our operating on automatic pilot, so to speak, as we do when we exhibit inattentional blindness or we ignore the "ground." Mindlessness represents an efficient and basically thought-free way of being in the world. The down side of mindlessness is that we can get caught up in seeing the world in rigid categories and so jump to incorrect conclusions. Our stereotypes and limiting beliefs are free to operate unexamined.

Another aspect of mindlessness is how we handle our mental commentary. Our minds continually chatter to us about what we are experiencing. If we mindlessly accept that chatter as the truth, we can end up drawing incorrect conclusions and acting on distorted information. If instead we recognize the chatter as hypotheses about what is happening, then we can test the hypothesis out and not be driven by it.

Journalist Malcolm Gladwell noted in *Blink: The Power of Thinking Without Thinking* that we may engage in "thin-slicing" or generalizing about a person or situation based on a tiny segment of information. Sometimes we rush to reach a conclusion without a lot of thought as a resistance to cognitive complexity. In the Myers-Briggs Type Indicator[11] personality test, this tendency is called "judging," or wanting premature closure. Gestalt psychologists[12] call it "functional fixedness," or accepting the obvious explanation and choosing not to think "outside the box."

Mindfulness, by contrast, is about seeing possibilities, remaining open to reframing experience, and developing new categories to explain our experience. It demands a more active role of us as the observer, acknowledging our context. Mindfulness involves paying attention to an experience in the moment without letting our concentration drift to what just happened, what will happen, or opinions about what is happening. As we pay attention in the moment, we do so without passing judgment—just notice and release. There is increasing evidence that we can learn mindfulness, and it can be useful in reducing stress and improving our overall health.[13]

> ### Exercise: Mindfulness
>
> Think about being with a friend or loved one who is distracted and preoccupied.
>
> Now think about being with that friend or loved one when they are really "present" with you so that you feel seen and heard.
>
> What is the difference you feel?

One important application of mindfulness is in reinterpreting what has happened. For example, imagine that you have gone into a coffee shop for a quick breakfast. Your server knocks over your cup of coffee while delivering your food and apologizes profusely. Your mindless reaction might be, "What a careless, clumsy person." If you took a second to look at the server (mindfully), you might notice that she has tired, haunted eyes and that her hands are trembling with fatigue. Now, instead of being focused on spilled coffee, you might focus sympathetically on her fatigue and distress. You might reframe your reaction as, "What a difficult situation she is in to have to work when she is obviously overtired and very worried," providing you with an excellent Compassionate Butterfly Shift opportunity.

Who We Notice

The whole point of the Butterfly Shift is to generate positive energy that can trigger a chain reaction or ripple effect. We probably already do this with people we know well, so our focus will be on initiating the Butterfly Shift with acquaintances or strangers. The concept of the Butterfly Shift is closely related to the growing movement of random acts of kindness.[14] The difference is the specific personal focus of the Butterfly Shift and the additional leverage that such focus can provide.

Exercise: Being Mindful

Think back on the last 24 hours and identify at least three people who helped you but whose help you took for granted:

1.
2.
3.

Why did you take it for granted?

What would need to shift for you to become mindful of such help?

For maximum effect, we want the Butterfly Shift interaction to be unexpected so that it has real emotional impact. Imagine that you have just prepared a special birthday meal and are getting thanked. That's not really unexpected, is it? In fact, you'd probably be a bit miffed if no one had noticed the special effort you made.

Now imagine that you have prepared a routine breakfast for your family in the midst of a hectic early morning. While usually your meal preparation is taken for granted, on this morning your partner says, "Thanks so much for getting breakfast—I know you have a busy day ahead, and I really appreciate it." How would you feel?

Since we are always choosing the focus of our attention, who will we single out for the special attention that is the hallmark of the Butterfly Shift? Every day we have a series of encounters with people who provide service to us. These may be people who help us in stores, provide service in our office, or who we talk with only by phone. They may be people we see repeatedly or people we encounter only once. We may know them by name, or they may be nameless to us. Each of these encounters offers a rich field of Butterfly Shift opportunities. When we thank someone who is used to getting a lot of recognition, the impact will not be the same as when we thank someone whose contribution is typically overlooked.

Choosing to notice service support workers serves an additional purpose. It is easy to notice people we think of as important, particularly if we hope that they will notice us in return. We feel somehow enhanced by that contact even though we have done nothing in particular to deserve any special recognition. But with routine service providers—at gas stations, grocery stores, dry cleaners, parking lots—we often feel entitled to be served, as if they owed us immediate attention. And, not surprisingly, what kind of treatment do we often get in return for that disrespect and projection of entitlement? Poor service.

Although we encounter many people in their work context, people are not defined by their jobs. Their contribution to a change process may be determined more by their emotional openness than by their job description. Research has shown that people who are emotionally more expressive are more likely to remember key social interactions than are those who are emotionally restrained even when the experience was positive.[15] So, we may wish to focus on people who are already in a neutral to positive mood since emotion is a key variable in the change process. For example, suppose you are in a store. You are more likely to make a Butterfly Shift difference with the service person who maintains eye contact and smiles as she rings up your purchases than with the person who rings up the sale correctly but with no emotional engagement.

> ### Exercise: Who We Notice
>
> Think for a moment about the past week. When you talked about your day, how often did you make comments about service encounters of the "you wouldn't believe how awful..." variety?
>
> How often did you make comments of the "you wouldn't believe how fantastic..." variety?
>
> What conclusions can you draw about who you are noticing?

Usually we choose to focus on the person right in front of us. But sometimes that person is not the one who is actually responsible for our positive experience. For example, suppose that your Elections Board has set up several early in-person voting locations, and you are pleased that this option is available because you will be away on the upcoming election day. You go to the voting location, which is staffed by volunteers. Were these volunteers responsible for this early voting option? Actually no. So you ask if there is anyone there from the Elections Board and are directed to a tired-looking, middle-aged man. You thank him explicitly, explaining why this extra voting opportunity is important to you and commenting that you realize it has taken extra effort to set up. He lights up. He's gotten a lot of complaints in his time, but no one has ever thanked him before. In this case, you have chosen the correct person to engage in the Butterfly Shift dynamic by seeking out a person responsible . . . and someone not used to getting compliments!

Exercise: Noticing Helpfulness

Think back over the past 48 hours. Can you identify a person who was helpful to you when they didn't need to be?

Did you notice that helpfulness at the time?

If so, what helped you notice? If not, why not?

One final consideration is the issue of chance encounters. Many of our examples are of structured settings where we expect to transact business. But what about the opportunities offered by someone who holds the elevator for you or who invites you to cut into a long line when you are in a hurry? These accidental meetings can be a marvelous chance to enact the Butterfly Shift since our appreciation will be unexpected.

What We Notice

Gestalt psychologists—psychologists who believe that the whole is greater than the sum of its parts—introduced the concept of "figure-ground" to explain how we organize bits of visual information. It describes the process of core elements coming together into a "figure" that stands out for us against a less noticeable "ground" or background.

In interpersonal interactions, we would describe what is most meaningful to us as the "figure" and what we tend to overlook as the "ground." For example, when we enter a clothing store, our attention will likely be drawn to the attractive clothing displays and any signs announcing sales (the "figure"). We are less likely to notice the sales clerks who are refolding clothes or returning them to the racks (the "ground"). For the purpose of the Butterfly Shift, it is precisely those "ground" supportive activities, usually taken for granted, that we want to notice.

One of our challenges in the noticing process is what psychologists call cognitive bias, the tendency to draw incorrect conclusions because of particular circumstances. Part of our mindless and often unconscious way of functioning is to simplify our lives by distorting the information we take in so that our worldview isn't threatened and we don't have to consider changing. Here are some common ways in which we exercise bias:

- Wishful thinking, including assuming the positive and forgetting any unpleasant memories.

- Limiting the data we consider by accepting the first option automatically or focusing only on information that supports our beliefs.

- Conforming to peer pressure or the expectations others have of someone in our role.

- Rejecting information simply because we dislike the person it comes from.

- Being influenced unduly either by the most recent or first information we get.

- Enhancing our sense of control by ignoring uncertainty or attributing our successes to our abilities and our failures to bad luck.

In determining what we will notice or focus on, we want to gather information that will allow us to be specific in our feedback once we move through noticing (Step One) and feeling (Step Two) and get to acting (Step Three). We will want to be able to evaluate the degree of effort that the other person is making and to acknowledge the most complicated aspect of the assistance we receive. The model developed to evaluate the quality of attention in patients with neurological problems[16] can help us do just that. The model distinguishes five kinds of attention, of increasing levels of complexity:

1. *Focused attention* — the ability to respond to specific cues or stimuli.

    For example, the service person answers your questions accurately about where to find items in a store.

2. *Sustained attention* — the ability to concentrate or maintain a consistent response during continuous and repetitive activity.

   For example, the service person remains pleasant throughout a complicated transaction that divides a detailed purchase between several credit cards.

3. *Selective attention* — the ability to filter or stay focused without becoming distracted when there is a lot going on.

   For example, the service person takes your food order successfully in a loud and busy restaurant.

4. *Alternating attention* — the ability to be mentally flexible and shift focus between tasks having different cognitive requirements.

   For example, you listen to a lecture while taking notes.

5. *Divided attention* — the ability to respond simultaneously to multiple tasks or demands.

   For example, the service person successfully serves several large parties in the same time period.

A related aspect of noticing is whether we focus on the positive or the negative, the glass half full or the glass half empty. Unless a situation is extremely unpleasant, we tend to notice the positive when focusing on the "figure." In the case of the "ground," we tend to remain mindless until something feels wrong, noticing only what is missing.

Of course, some people have a propensity to be pessimistic, always noticing what is wrong. Research shows that many people are more likely to focus on complaints than on satisfaction and to follow up by telling their complaints to at least eight other people.[17]

> ### Exercise: Optimism and Pessimism
>
> Think about yourself for a moment. Are you naturally optimistic and able to find a silver lining in any situation?
>
> Or are you naturally pessimistic and focused on anticipating what could go wrong?
>
> If you are a natural pessimist, what would help you to find the positive in a situation so that you could engage in a Butterfly Shift?

Leveraging Our Ability to Focus

Until we think about it, we may be unaware of how much of our routine life goes on without our conscious direction. In particular, we typically take for granted a wide range of supportive activities that help make our lives run smoothly. We usually become mindful only when our expectations are not being met, when there is a deficit or something negative has happened. This then generates an overall sense of irritation and frustration.

If the Butterfly Shift is to be effective as a mini-immersion experience, we will need to shift what we notice and give special attention to. We will need to become mindful of the positive happening around us instead of being triggered into mindfulness only by things that are not going our way. We will need to shift our focus from an absorption with ourselves and our needs as the "figure" and notice instead the helpful supportive activities that we usually assume as "ground" and take for granted.

The daily Butterfly Shift practice offers us the opportunity to participate meaningfully in our own lives from a positive, rather than a negative, perspective. In choosing what we will notice, we can mindfully create a warm and generous ambience. By noticing helpful actions, we gradually fill up our awareness with kindness and thanksgiving, crowding out hurt and despair. And, through our focus, we become a joyous participant in our bountiful cosmos.

FIVE

Noticing — Step One

What we notice shapes our reality — literally! So our first step in the Butterfly Shift mini-immersion is to practice changing our focus, shifting our attention from ourselves to others. Instead of approaching situations from a "what's in it for me" perspective, we need to adopt the attitude of "walking a mile in another's shoes." This shift from self-preoccupation to appreciation of others is by itself dramatic in its ramifications. It is key to our being "present" with others and with ourselves.

We will be most successful in the long term if, each day, we look with anticipation for someone to notice . . . someone different than the day before. That way we expand the range of our awareness and the degree to which we are "in the flow" instead of being self-absorbed and feeling entitled.

Of course, the great thing about a daily practice is that we get plenty of chances to experiment — seven times each week. So we can try out different noticing strategies and see which ones are most effective for us. And we can practice targeting different types of people for the Butterfly Shift. As part of our "harmlessness mini-immersion," we want to expand our repertoire of harmless engagement.

Ways to Improve Your Noticing Skills

We already know that noticing is not automatic and improves with practice. Here are five strategies that can help you fine tune the skills you need for Step One of the Butterfly Shift:

1. *Notice your context*

    To avoid inattention and "cognitive capture," put away your cell phone and iPod and focus on where you are right now. Be mindful and mentally present with the person in the moment. Be aware of the small details of your interactions with people who are helping you. And be alert to your assumptions because they will influence what you notice.[1]

2. *Notice in a flexible manner*

    To avoid being limited by assumptions, cultivate a sense of openness to, or curiosity about, what you might experience. Be alert to cultural filters or to drawing conclusions too quickly (thin-slicing).

3. *Notice respectfully and engagingly*

    Since it is all too easy to disregard service persons who are being helpful, imagine that other person as a favorite cousin—someone not seen frequently but still valued. The warmth and concern you feel will convey itself in your nonverbal communication, especially if you try to link it to the other person's needs and priorities.

4. *Notice the "ground" and not just the "figure"*

    To avoid taking people and assistance for granted, pay attention to people who are not immediately similar to you. When you find yourself focusing on the negative, try reframing what you are noticing from a positive perspective.

5. *Notice expressively*

   To increase the chances that the other person will participate in the Butterfly Shift dynamic with you, be emotionally present and engage in a manner that is actively positive. In other words, control and project a positive emotional state.

Focus and the Type of Shift

Of course, who you choose does depend on the type of Butterfly Shift you plan to employ. Here are some pointers:

1. For the *Compassionate Shift*:

   Here the circumstances are more important than the amount of help you receive. You will want to pay particular attention to persons who are trying to be pleasant and helpful under difficult circumstances, such as feeling very tired or distressed or overwhelmed.

2. For the *Grateful Shift*:

   You will want to focus on the "invisible" helpers who truly go the extra mile in helping, or being supportive of, you.

3. For the *Joyous Shift*:

   You will want to select someone who looks like they could use an encouraging word but is not expecting it.

Choosing Your Focus

Who might we choose to notice as part of the Butterfly Shift dynamic? Remember that in each instance we want to maximize the likelihood that the Butterfly Shift interaction will result in an energetic shift—for ourselves and for the other person.

Step One is about selecting someone for your focus based on the type of Butterfly Shift you've chosen—Compassionate, Grateful, or Joyous. Depending on the circumstance, you could select someone based on their level of engagement in being helpful to you, their emotional openness to the Butterfly Shift dynamic, their relationship or status with others, or how responsive they are likely to be. Here are descriptions of eight possible types of persons you might choose:

Level of Engagement

1. Choose a person who is directly responsible for the action or encounter that made a positive difference to you. They would be the one to replicate what created your sense of satisfaction. In many cases, this will be the specific person with whom you interact. But sometimes it will be someone in a decision-making capacity who you need to seek out.

2. Choose a person who has made an effort beyond what was absolutely necessary so that you are able to praise them for something unusual.

Openness

3. Choose a person who is emotionally expressive. They are more likely to interact willingly with you and are more likely to engage with others, creating an emotional ripple effect.

4. Choose a person who is already in a neutral to positive mood so that they are likely to be open to noticing and engaging with you.

Interpersonal Considerations

5. Choose an older person over a youth in order to be respectful of life experience. In many multicultural set-

tings, this choice would be considered to be culturally sensitive.

6. Choose a person whom you've observed as including their colleagues and other customers in their interactions. They are more likely to pass the energy along.

Responsiveness

7. Choose a person who seems receptive to an interaction — someone who will maintain eye contact, who will respond to a conversational gambit, who will listen so that you have a good chance of success in interacting with them.

8. Choose a person who does not seem to be singled out for acknowledgement regularly, and is likely to appreciate the attention.

Exercise: Choosing Your Focus

Of the eight types of people described above, which type would you find it easiest to compliment, and why?

Which type would you find it hardest to compliment, and why?

What could you do to expand the types of people that you choose to compliment as part of the Butterfly Shift?

Ensuring the Success of Step One

When all is said and done, we can only change ourselves. Any influence we might have on others comes from shifting how we behave. Each day Step One sets the stage for the Butterfly Shift we will select. As we notice the person trying against great odds to be helpful or the person who goes out of their way without expecting recognition for it, we begin to engage differently with our community. And we start to experience what living harmlessly could be like.

SIX

Leveraging Emotions

Emotional energy influences us throughout the day. When someone around us is happy and cheerful, we usually blossom like a flower unfurling in the sunlight. When someone is sad or upset, we sense it even before they say something. When a driver honks their horn or swerves past us with finger raised, we begin to get irate.

Some of our responsiveness we can account for by the small behavioral cues we pick up—the smile, the hunched shoulders, the raised finger (not too subtle!). But not all of it. We actually have phrases for this emotional influence. We say, "You could cut the tension with a knife" or "The room hummed with energy."

The Importance of Emotions

Metaphysics teaches us that "energy follows thought,"[1] that what we conceptualize is what we will manifest. We experience many offshoots of this belief—for example, the teachings of the power of positive thinking, or sport psychology's emphasis on the visualization of success. And they are all partially correct.

But if the primary thing that mattered was thought, then all kinds of things would manifest just because we thought about them. We could stop at Step One of the Butterfly Shift and expect simply noticing or thinking about an action to make all the difference in our mini-immersion experience.

Emotion is a critical catalyst for any change, particularly strong emotion. Think of emotion as the flame that warms our energy field, making the energy more pliable, more dynamic. Neuroscientist Candace Pert has asserted, "Emotions are the connectors, flowing between individuals, moving among us as empathy, compassion, sorrow, and joy. . . . It is a scientific fact that we can feel what others feel. The oneness of all life [the Principle of Interconnectivity] is based on this simple reality: Our molecules of emotion are all vibrating together."[2] Because of the "all-nourishing abyss"[3] through which we are all connected, our behavior—even the simple act of observing how we feel—affects our shared energy field.

But emotions often seem messy and unsettling. In an era where scientific objectivity and rationality are still highly valued as part of the deterministic model of Newtonian physics, emotions seem to get in the way. We describe people as "overly" emotional or "too" emotional. One of the ongoing gender role tensions is the assignment of feelings as "feminine" and mental thoughts as "masculine," so that the person being labeled as "too" emotional is usually female. If we stop to examine this classic dichotomy, though, it is actually very odd.

Where do feelings, our conscious experience of emotion, come from anyway? They spring from thoughts, from our explanations to ourselves (verbal or nonverbal) of what is going on. These explanations influence how we feel, how we respond, whether or not we personalize another's actions.

On the mental side of the dichotomy, what gives thoughts their potency? Emotional energy. Thoughts by themselves are arid. When we describe someone as an "intellectual," this is usually not a warm, cozy description. We are alluding to thought without emotion, head without heart.

The philosopher Robert Solomon describes an emotion as "not only an interpretation of our world but a projection into its future, filled with desires which sometimes become intentions and commitments."[4] So emotions serve as motivators or predispositions to act. They give meaning to what we experience.

Our intuitive sense is tied to our emotions. "Emotional energy works at a faster speed than the speed of thought. This is because the feeling world operates at a higher speed than the mind. Scientists have repeatedly confirmed that our emotional reactions show up in brain activity before we even have time to think. We evaluate everything emotionally *as* we perceive it. We think about it *afterward*."[5]

> **Exercise: Making Decisions**
>
> Pick an everyday decision you are about to make. First say, "I feel like . . . [whatever the decision is]."
>
> Then say, "I think I will . . . [whatever the decision is]."
>
> Do you experience any difference in the two approaches? If so, what difference?

Research shows that, in certain situations, we actually make judgments faster and with greater confidence based on our feelings than based on a mental assessment.[6] In fact, there is growing evidence that decisions based on thought and decisions based on emotions come from two separate and partially independent internal systems.[7]

Emotions determine our relationships with others and our environment. When we feel positively, we move towards others. When we feel negatively, we move away or push others away. So emotions mediate the types of relationships we establish and sustain.[8] Other people are influenced by, and respond to, the emotions we express and then we react to the response of others, so we can think of emotions as occurring in cycles.[9]

We display emotions not only through words but also through our nonverbal behavior. One component is known as paralanguage—or the pitch and pacing and volume of how we speak. For example, we pick up the excitement of another person in part from their rapid, almost breathless, speech. Another component is our body language—facial expressions, hand gestures, posture, eye contact. There are certain innate facial expressions for emotions that researchers have shown are the same for sighted persons and persons who have been blind from birth and that seem to be the same worldwide.[10]

Developing Our Emotional Muscle

Emotional expression is similar to a physical muscle—it gets stronger, more flexible, and more nuanced the more we practice it. We are not talking here about "being emotional"— that state of feeling out of control. Rather, we are talking about the ability to regulate our own emotions and focus them to achieve goals. This is the self-management component of emotional intelligence, a concept made popular by Daniel Goleman.[11]

If we truly want to leverage our emotions as part of the Butterfly Shift dynamic, we need to understand and strengthen three aspects of our emotional expression: *valence*, or positivity; *range*, or breadth; and *intensity*, or strength.

Emotional Valence

We call whether an emotion is positive or negative its *valence*. Our negative emotions—like anger or fear—come with a sense of aversion or repulsion, of wanting to disown the experience we are having. They constrict us and close us off to collaborating with others.

Negative emotions are processed in different parts of the brain than positive ones,[12] which may explain why we recall negative emotions more vividly than positive ones. Not only are we generally more sensitive to negative emotions, but females in particular react strongly to even mildly negative facial emotions.[13]

Anger and fear are the emotions most frequently researched and often the most intense emotions people feel on a regular basis. Anger is the more energizing of the two, and we may engage in anger in order not to feel frightened. When customers of service firms were asked about their emotions, anger was the most frequently mentioned.[14] Although anger comes from our own sense of being thwarted, we usually direct it at others—"it's all his fault."

All negative emotions result in a kind of tunnel vision that limits the options we see and the degree to which we can be creative. The stereotype of emotional torment being required for artistic creativity is simply not true.

By contrast, positive emotions are the ones that draw us in and engage us more closely. They open our hearts and create an energetic field ripe for change. So, for the purposes of the Butterfly Shift dynamic and the intensity of the mini-immersion, we want to work with positive emotions.

We actually know much less about positive emotions than about negative ones. Historically, psychologists have considered positive emotions "lightweight" and of little real importance. It is only since the late 1990s that researchers have started to recognize the ability of positive emotions to open the mind to multiple possibilities and to creative thinking.[15] There are not only several Positive Psychology Centers, but there is also a *Journal of Positive Psychology*, which would have been unthinkable as recently as the 1980s.

What we do know about positive emotional experiences is that they are significantly less likely to result in false memories than are negative experiences—in other words, positive emotions match well with our actual experience.[16] In fact, positive feelings can enhance the vividness and accuracy of our memory, giving it a kind of glow.[17]

> **Exercise: Momentum of Emotions**
>
> Test the momentum of emotions for yourself during the next 24 hours.
>
> Notice a time when you feel really happy, and try to pull back from people.
>
> Notice a time when you feel upset, and try to feel open towards others.
>
> Did these attempts feel natural or like you were trying to move against the grain?

On a practical level, our growing reliance on mobile phones and similar personal communication devices brings with it positive feelings of intimacy and connectedness with friends and family.[18] This illustrates the multidimensionality of our emotional experience. Of course, we are also facing challenges in how we manage our private emotional life in a very public context.

Fortunately for the Butterfly Shift dynamic that we are interested in, positive emotions are more likely to motivate people to behave in ways that are constructive and helpful.[19] Happiness, in particular, broadens our awareness of options and our ability to "think outside the box." Our brains recognize happiness more quickly than negative emotions. Feelings of anger can be reduced or even completely disrupted by providing a reason for the person to feel happy.[20]

This last research finding is particularly important for implementing the Butterfly Shift as a mini-immersion experience. While persons who are already in a positive mood may be more receptive to us, by projecting a strong positive emotion, we can help others become positive.

Emotional Range

Emotional *range* refers to our repertoire of emotions, or which emotions we can express. It determines the degree of connection we are likely to make with another person. Think of *range* as an emotional palette. Let's say that the only color we can express is yellow. That may work well in connecting with people who function in the orange or even green color ranges as both those colors contain yellow, but we would be out of luck trying to communicate with someone in the purple range as purple does not contain yellow.

Keep in mind that our purpose in the Butterfly Shift dynamic is to create and sustain a positive interaction with another person. So we have two challenges. One is to identify the positive emotions that we are comfortable expressing. The other is to understand the range of positive emotions with which the other person may resonate. Remember—the issue ultimately is not the emotion we are most familiar with but the emotion that will best connect us to the other person.

The psychologist Robert Plutchik has proposed a psycho-evolutionary model of emotion that has become widely accepted.[21] It is structured around eight basic emotions: anger, disgust, sadness, surprise, fear, trust, joy, and anticipation. Notice the heavy weighting towards the negative emotions! So our first task is to become aware of an expanded range of positive emotions, such as those listed in the table of "Examples of Positive Emotions" below.

Examples of Positive Emotions

affectionate	amused	blissful
appreciative	cheerful	delighted
benevolent	generous	empathic
compassionate	grateful	exhilarated
considerate	happy	ecstatic
content	hopeful	elated
gentle	humbled	gleeful
glad	kind	hilarious
patient	magnanimous	joyful
sympathetic	playful	jubilant
tender	pleased	overjoyed
thoughtful	respectful	passionate
understanding	serene	rapturous
warm	thankful	thrilled

While not all researchers agree on precisely which emotions are basic, there is evidence that at least anger, fear, disgust, sadness, and surprise are emotions expressed by everyone. And both laughter and crying are universally understood.[22]

Many of us have more emotional range when it comes to negative emotions as compared with positive emotions. Indeed, some of us may find negative emotions—like anger—more comfortable and familiar than positive emotions.

Leveraging Emotions 99

What do we typically do with our negative emotions? Customer service research shows that we are four times more likely to complain than praise, and ten times more likely to tell others about a negative experience than about a positive experience.

> **Exercise: Positive and Negative Emotions**
>
> In the last 48 hours, which positive emotions from the list above have you experienced?
>
> Now, which negative emotions—including sadness, irritation, fear, impatience, frustration, anger, rage, fury—have you experienced?
>
> Which group—positive or negative—occupied the majority of your awareness?
>
> Regarding the positive emotions that you feel less frequently, how could you make them a regular part of your repertoire?

Our reality is that negative emotions are more frequent and more familiar. Reflecting back on the dynamic of mindfulness, we remember that mindfulness is typically triggered by negative emotions—by something going wrong. So we are socialized into a negatively-focused emotional frame of reference as part of our norm of harmfulness.

In order to immerse ourselves in the possibilities facilitated by the Butterfly Shift, we need to develop experience with a wide range of positive emotions. And we need to practice letting a positive response be our emotional default.

Emotional Intensity

Emotional *intensity*, or depth of emotion, is what creates resonance or energetic impact. Imagine two different scenarios where you connect with someone you have really missed. In the first scenario, the other person says, "Oh hi, how're you doing." In the second scenario, the other person says, "How fantastic to hear from you! I've really missed you!" The difference in response is intensity. The difference in your engagement with the other person will be marked.

Now look back at the list of positive emotions in the "Examples of Positive Emotions" table above. They are in three columns, with the first column being the least intense and the third column being the most intense. Are the positive emotions you experience primarily from one column rather than the other two? If so, which one? To be most effective in facilitating change, you will want to draw on very intense positive emotions (the far right column).

Exercise: More Intense Positive Emotions

Pick an emotion in the "least intense" column that you commonly experience.

What very intense positive emotion from the third column could you substitute?

How would you do that in an authentic manner?

One of the challenges we face is that emotions like anger or fear or disgust are usually intensely felt. If we want a rush of the energy that comes with an intense feeling, we are used to getting it from a negative emotion.

The most common intense positive emotion we experience is falling in love—which we don't do every day! Another type of intense positive emotion is blissfulness, which we associate with unusual circumstances such as peak experiences. While we talk about being joyous, it is unusual for people to sustain an intense positive emotion. It takes practice.

Some of us may fear emotional intensity. If we feel intensely and that intensity is not matched by the object of our feelings, we may fear that we look foolish. Or we may be uncomfortable with the intensity itself. Or we may believe that very intense emotion can unduly influence another person. Remember that we will always have an effect on others, regardless of the intensity of our emotion. What will be apparent is whether we are expressing our true feeling or whether we are faking a feeling. And each person always has a choice about how to respond to an emotional message. So the issue is not whether or not to have an effect, but rather the type of effect we wish to have.

In the research literature, the term "arousal" is often used to refer to intensity. We are already aware that feelings aroused by fear help us focus our attention.[23] With any intense emotion, we become more selective in what we pay attention to and, as a result, we usually have a more accurate memory of what has happened.[24] When intense emotions are involved, not only does our memory become more accurate, but it improves over time—we consolidate what we remember.[25]

In everyday life, we find that people are most comfortable with those whose emotional expression shares the same intensity as their own. But, even when intensity levels match, the gender of the person may provide an opportunity for misunderstanding. Research shows that, when the intensity of emotion being expressed is actually the same, men will be seen as expressing stronger anger and women will be seen as expressing the most intense happiness.[26] So, if you are male and wish to express warmth and happiness, you will need to crank up your intensity.

Type of Emotion Linked to Type of Shift

Shift	Gentle	Moderate	Intense
Compassionate	benevolent *compassionate* considerate gentle patient sympathetic tender thoughtful understanding warm	generous kind	empathic
Grateful	appreciative	*grateful* hopeful humbled magnanimous pleased respectful thankful	delighted
Joyous	affectionate content glad	amused cheerful happy playful serene	blissful exhilarated ecstatic elated gleeful hilarious *joyous* jubilant overjoyed passionate rapturous thrilled

Now think about the three types of Butterfly Shifts in terms of their overall intensity—Compassionate, Grateful, Joyous. Which emotions would you think of as linked to each type of shift? The table above shows an interesting pattern. Most of the emotional nuances related to *compassion* are gentle emotions, of relatively low intensity. Those related to *gratitude* are of a stronger, but still moderate, intensity. By contrast, most of the emotions related to *joy* are of very strong intensity. And the nuances shown above are made more or less intense by our accompanying nonverbal communication.

The intensity of an emotion is what gives it the power to help us manifest intentions. It is our emotions that regulate what we experience as reality. The trick in the Butterfly Shift is to be able to feel intensely while also detaching from the outcome. If we use a musical analogy, we can think of a change in frequency or pitch as resulting in a change in valance from negative to positive, a change in amplitude resulting in a change in range, and a change in volume resulting in a change in intensity.

Emotional Contagion

One other aspect of our emotional experience that is relevant to the Butterfly Shift is emotional contagion. Just as we tend to become ill around people who are physically contagious with a disease, we pick up the emotional residue of people who are expressing an emotion particularly intensely.[27]

"Emotions serve to focus our attention on aspects of the world that help us thrive. They provide information about our interior world and about our relationships. For this survival function to operate optimally, we are highly sensitive to emotional signals in the environment. One person's emotions are affected by others'."[28] Over time we become more and more adept at picking up and reading the feelings of others.[29]

Unfortunately, we are more likely to pick up other people's moods ("mood contagion") when they are negative than when they are positive. We have terms for this dynamic like "hey, you are bringing me down." The contagiousness of emotions makes it imperative that we take responsibility for our own emotional output. If we feel negatively, we create a kind of emotional pollution that we need to contain and dissipate before it affects others. Conversely, we need to be responsible for not taking on others' emotions as our own.

Exercise: Emotional Contagion

Pick a time in the recent past when you were feeling good and came into contact with someone you knew who was very distressed.

How long were you able to stay in your positive emotional space without being affected by the other person?

If you were affected, why?

If you were able to stay unaffected, what helped?

How did you feel as a result of either being or not being affected?

To benefit from the Butterfly Shift, we will want to ensure that we are radiating positive emotion for others to pick up. Fortunately, not only our positive emotion but also our positive intentions can be sensed by others and serve to enhance their positive mood.[30]

The Power of Emotions

Japanese scientist Masara Emoto has generated stunning and controversial pictures of water crystals, showing that thoughts and feelings do affect our physical reality.[31] When positive emotions are focused on water, the resulting crystals are brilliant, complex, and colorful. When water is exposed to negative emotions, its crystals are incomplete and asymmetrical, in drab colors. If we consider that humans are at least 70 percent water, the potential impact of emotional valence is astounding.

Indeed, "emotions are us" says Pert, author of *Molecules of Emotion*. Energy itself is neutral. We provide its "charge" through the neurochemical activities that underlie, or produce, emotion. Emotion-charged energy is what flows between us, connecting us. What alternative medicine knows as *chi* is in actuality the free flow of information carried by the biochemicals of emotions — neuropeptides and their receptors. Pert provides a beautiful analogy of the relationship between the physical and the emotional: "Peptides serve to weave the body's organs and systems into a single web that reacts to both internal and external environmental changes with complex, subtly orchestrated response. Peptides are the sheet music containing the notes, phrases, and rhythms that allow the orchestra — your body — to play as an integrated entity. And the music that results is the tone or feeling that you experience subjectively as your emotions."[32]

There is a growing body of research showing that our emotional energy affects our health. While positive emotions may not cure us once we are ill, they do definitely protect us against becoming sick.[33] Research has shown that providing emotional support to others, including caretaking for a spouse, results in a 30 to 60 percent reduction in mortality risk and so lengthens our life.[34] Helping others truly does help ourselves, as well as precipitating change.

The Heart and the Brain

"The heart is not just a muscle. It's also a sensory organ and a sophisticated information processing center. The heart actually has its own nervous system, which gives it the ability to sense, learn, remember and make functional decisions independent of the brain. . . . The heart is part of the emotional system and . . . plays an important role in how we feel and think. . . . Emotions are reflected in the patterns of our heart rhythms. . . . Positive emotions like appreciation, care, compassion and love lead to a more ordered and coherent heart rhythm pattern."[35]

We may not be used to thinking of the heart as a powerful generator of electromagnetic energy, but the Institute of HeartMath has shown that its magnetic field is 5,000 times stronger than that of the brain.[36] The heart rhythm coherence produced by positive emotions helps us perform better cognitively. Of particular relevance to the Butterfly Shift is the fact that "when two people are at a conversational distance, the electromagnetic signal generated by one person's heart can influence the other person's brain rhythms."[37]

Emotions are what trigger the choice underlying decisions. In fact, we get into major trouble when our reflexive choice-making function goes awry. Science writer Jonah Lehrer[38] has provided us with a stimulating overview of what happens on the neurophysiological level when the link to our emotional sense is severed. If we are unable to feel emotion, we become literally unable to make decisions. Our brain depends on visceral emotions as a rapid response system—much quicker than cortical analysis—to set the stage for choices and decisions.

Leveraging Specific Emotions

Each type of Butterfly Shift—the Compassionate Shift, the Grateful Shift, the Joyous Shift—is linked to a specific family of emotions. In general, we want to focus on intense versions of the positively-valenced emotions indicated in grayed cells of the "Type of Emotion Linked to Type of Shift" table above. In order to use these emotions effectively, we need to understand each "lead" emotion and then how to expand our range and intensity within that group of emotions.

Compassion

Compassion is a profound emotional response to the pain and suffering of others. It is more intense than empathy and is the fundamental or "lead" emotion in our first type of Butterfly Shift—the Compassionate Shift. Compassion is considered one of the great virtues in major religious traditions. It is a direct expression of the Principle of Interconnectivity—we are all interconnected and so your pain is my pain.

The Dalai Lama has been quoted as saying, "If you want others to be happy, practice compassion. If you want to be happy, practice compassion." The Buddhist monk Bhikkhu Bodhi states that compassion "supplies the complement to loving-kindness: whereas loving-kindness has the characteristic of wishing for the happiness and welfare of others, compassion has the characteristic of wishing that others be free from suffering."[39] In Islam, one of the purposes of fasting during Ramadan is to create compassion for those not fortunate enough to have enough to eat and sensitivity to the suffering of others. As one Jewish rabbi has said, "Kindness gives to another. Compassion knows no 'other.'" The Christian theologian Thomas Merton has said, "Compassion is the keen awareness of the interdependence of all things."

When Karen Armstrong won the 2008 TED prize,[40] her "One Wish to Change the World" was to develop a global Charter for Compassion. The purpose of the Charter (presented below) is to leverage compassion, as a tenet of all major religions, to build common ground and a harmonious global community.

Charter for Compassion[41]

The principle of compassion lies at the heart of all religious, ethical and spiritual traditions, calling us always to treat all others as we wish to be treated ourselves. Compassion impels us to work tirelessly to alleviate the suffering of our fellow creatures, to dethrone ourselves from the centre of our world and put another there, and to honor the inviolable sanctity of every single human being, treating everybody, without exception, with absolute justice, equity and respect.

It is also necessary in both public and private life to refrain consistently and empathically from inflicting pain. To act or speak violently out of spite, chauvinism, or self-interest, to impoverish, exploit or deny basic rights to anybody, and to incite hatred by denigrating others—even our enemies—is a denial of our common humanity. We acknowledge that we have failed to live compassionately and that some have even increased the sum of human misery in the name of religion.

We therefore call upon all men and women ~ to restore compassion to the centre of morality and religion ~ to return to the ancient principle that any interpretation of scripture that breeds violence, hatred or disdain is illegitimate ~ to ensure that youth are given accurate and respectful information about other traditions, religions and cultures ~ to encourage a positive appreciation of cultural and religious diversity ~ to cultivate an informed empathy with the suffering of all human beings—even those regarded as enemies.

We urgently need to make compassion a clear, luminous and dynamic force in our polarized world. Rooted in a principled

> determination to transcend selfishness, compassion can break down political, dogmatic, ideological and religious boundaries. Born of our deep interdependence, compassion is essential to human relationships and to a fulfilled humanity. It is the path to enlightenment, and indispensible to the creation of a just economy and a peaceful global community.

Scientific studies suggest that there are a number of physical benefits to the practice of compassion including the increased production of DHEA, which counteracts the aging process, and a reduction in the "stress hormone" cortisol. The practice of compassion increases our capacity to care or empathize with others.[42]

Compassion is central to the first type of Butterfly Shift because of its importance in helping us make the shift from an ethic of harmfulness to one of harmlessness. When we truly identify with the pain and suffering of others, it becomes impossible for us to be violent towards them. So compassion fixes our attention lovingly on the other.

Gratitude

Gratitude — thankfulness or appreciation — acknowledges a benefit that we have received or will receive. In contrast to compassion, gratitude focuses on our own situation and how others have played a role in shaping our experience. The practice of gratitude is the focal point of the second type of Butterfly Shift — the Grateful Shift.

Gratitude is usually thought of as an emotion that occurs after we receive help, but we can also experience it in anticipation of being helped. We are most likely to feel gratitude if we see the help we receive as (a) valuable to us, (b) requiring effort on the part of the giver of the help, and (c) being given with no strings attached and no ulterior motives.[43]

Gratitude may be uniquely important in determining our wellbeing, as research shows that people who are more grateful cope better with a life transition and are more willing to be helpful to others.[44] Research has linked the experience of gratitude to being happier, less depressed, less stressed, and more satisfied with life and social relationships.[45] As with compassion, we are the indirect beneficiaries of expressing gratitude towards others.

Grateful people also report a heightened sense of control over their environments, a greater sense of life purpose, and increased self acceptance.[46] Grateful people have more positive ways of coping with the difficulties they experience in life, being more likely to seek support from other people and to grow from the experience rather than denying that there is a problem or blaming others.[47]

Gratitude is not the same as a sense of indebtedness, which implies an obligation of repayment.[48] Instead of wanting to avoid that reminder of indebtedness, gratitude motivates persons to want to develop a stronger relationship with the person(s) who helped them.[49]

Joy

Joy, the most intense of the positive emotions, is the foundation for our third type of Butterfly Shift—the Joyous Shift. In this type of shift, we are leveraging our own joy to stimulate a joyous resonance in others.

Metaphysics hypothesizes that "joy is the strong basic note of our particular solar system."[50] If you doubt the fundamental nature of joy, think about young healthy infants. They are spontaneously joyous, and it is difficult to be with such an infant without smiling yourself.

While we still know relatively little about the emotion of joy, its less intense sibling, happiness, has been well researched through a growing network of "positive psychology" centers.

Happiness can be defined as a feeling of contentment, satisfaction, pleasure, or joy. It is linked to a focus on *being*, no having. Various quality of life measures, such as Bhutan's Gross Happiness Index, are being developed as part of the growing economics of happiness.[51]

One of the characteristics of happiness is that it radiates. When we feel happy, that energy is very apparent to others. Happiness produces a sense of hope that helps us grow and thrive because it offers a sense of possibilities.

In his book *Authentic Happiness*, psychologist Martin Seligman describes happiness as consisting of positive emotions related to the past, present, and future. Along this time line, emotions range from satisfaction, contentment, pride, and serenity to optimism, hope, and trust. Seligman maintains that our sense of fulfillment comes from exercising our unique strengths and virtues in a purpose greater than our own immediate goals. Perhaps most relevant to our Butterfly Shift, he has documented "learned optimism" — or the ability to alter our outlook on life by reframing our circumstances to reinforce an intensely positive outlook.

Maximizing Benefits From Emotions

One of the consequences of engaging in the Butterfly Shift dynamic is that we have an opportunity to change our emotional engagement with the world. We have seen that, aside from the experience of falling in love, our most intense emotions tend to be negative rather than positive. The Butterfly Shift focuses us instead on building our positive emotion "muscle" by increasing the frequency, range, and intensity of the positive emotions we experience.

Emotions are powerful, and their impact reflects our interconnectedness. They determine what we attend to and how we choose to act. "Emotions are constantly regulating what we experience as 'reality.'"[52] We are more likely to recall positive events when we are in a good mood and are also more likely to be helpful and altruistic. Each of the critical positive emotions critical to the mini-immersion experience of the Butterfly Shift—compassion, gratitude, joy—take practice.

SEVEN

Feeling — Step Two

What we notice is critical. The emotional ambience we create is just as critical. In fact, feelings are the change component most often overlooked. To be effective as a mini-immersion experience, the Butterfly Shift requires a strong positive emotional expression. Step Two is about exercising our positive emotion muscle.

Ways to Improve Your Feeling Expression

We have seen in Chapter Six that we are still relatively inexperienced at expressing positive emotions as compared with negative ones. As much as 70 percent of our energy comes from nonphysical sources, such as our emotions, so it is important that we practice positive ways to energize ourselves. For any who have difficulty getting started, Laughter Yoga is a technique that can be very helpful.[1] There are now more than 6,000 Social Laughter Clubs in 60 countries facilitating the practice of laughter.

There are also some strategies that you can practice in a wide variety of situations, not just during a Butterfly Shift, to strengthen your feeling expression. These can help you engage more actively with the positive emotional experience that is a critical part of an ethic of harmlessness:

1. *Focus on the positive, not the negative*

    Find the positive component in any situation, even if it is only what you can learn from the situation.

2. *Increase the range of your positive emotions*

   Practice experiencing each of the emotions listed in the "Examples of Positive Emotions" table in Chapter Six at least once a week. Be particularly mindful of practicing the nuances of positive emotions you don't often express.

3. *Increase your control of your emotional state*

   Consciously refocus away from your own concerns and preoccupations and display a positive emotional expression if you find yourself drifting towards more familiar negative or neutral emotions.

4. *Increase the intensity of your positive emotions*

   Begin at a lower intensity with each positive emotion, then increase that intensity as much as you can, and then bring the intensity back down to your starting point. Do this until you can express a specific intensity at will. This is a bit like practicing musical scales, going from a low note to a high note and back down. You can use tone of voice and facial expression to increase intensity.

5. *Match your emotional intensity to others' intensity*

   Mirror the intensity of the person with whom you are seeking rapport in order to display empathy, which facilitates the Butterfly Shift process. You don't necessarily have to mirror the emotion itself. If the other person is feeling sad, for example, you could express the same intensity of compassion.

6. *Use a smile to generate a positive feeling*

   Act as though you feel positive even if you don't. Research has shown that creating the facial expression of a particular emotion evokes the feeling.[2] This finding underscores the fact we can control our emotional state.

One of the methods that we often use to express positive emotion is through humor. Be aware that much of what is humorous is culture-specific and can be misunderstood. If you are with persons from other cultures, you might want to avoid using humor. There is also the danger, in using humor, that you may be heard as being sarcastic rather than positive.

Emotion and the Type of Shift

For Step Two of the Butterfly Shift, the emotion you choose to express will of course be in the family of the one that precisely matches the type of shift you have chosen—the Compassionate or the Grateful or the Joyous group of emotions. You have flexibility, though, regarding which emotion within that family you choose and the intensity with which you express it. Here are some guidelines for maximizing the emotional power of each type of shift.

1. For the *Compassionate Shift*:

    You can cultivate the expression of compassion by:

    a) Being gentle and understanding with yourself. You are likely to treat others the way you treat yourself.

    b) Focusing on what you have in common with the other through assertions such as the following:

    - "This person has felt despair, just like me."
    - "This person is trying to avoid suffering, just like me."

- "This person is learning about life, just like me."
c) Projecting yourself into the place of the other person and determining what you would want another to desire for you.

d) Being willing to forgive those who have clearly mistreated you or ones you love by recognizing that we all make mistakes.

There is one caveat to the expression of compassion. You need to be able to maintain your own energetic boundaries and not absorb the pain and frustration of others. Known as "compassion fatigue," this dynamic refers to the lessening of compassion over time because of prolonged exposure to victims of trauma.

What happens is that the sufferer of compassion fatigue becomes swamped by the emotions of others instead of maintaining their own emotional integrity. On an energetic level, we can think of this as the person's aura or energetic boundary being too porous and letting too much of someone else's energy through. One of the simplest ways to strengthen our auric boundary is to visualize ourselves surrounded by a bubble of white light, with an outer "edge" that is like the semipermeable membrane of a cell. Its function is to let in the energy that is healthy for us and to keep out any energy that is not healthy.[3]

2. For the *Grateful Shift*:

You can cultivate the feeling of gratitude by following strategies such as:

a) Establishing a practice of listing the people and things for which you are grateful as part of your daily meditation or reflection.

b) Finding a "gratitude buddy" — someone with whom you regularly talk about what you are each grateful for.

c) Noticing how you feel when you are grateful — the overall sense of wellbeing, the positive shift in your mood.

If you have difficulty with expressing gratitude, you may find that *A Network for Grateful Living* has some resources that are useful to you.[4]

Be aware that there are gender differences in how people respond to an expression of gratitude.[5] Women tend to be more receptive to gratitude without feeling any sense of obligation. Men, on the other hand, are more critical of gratitude and resist feeling or expressing it.

3. For the *Joyous Shift*:

Joy is the most dramatic of the emotions being targeted in the Butterfly Shift, and its expression is often accompanied by expansive physical gestures. You can enhance your likelihood of experiencing and expressing joy by:

a) Practicing exaggerating and dramatizing your expression of joy so that, by comparison, a usual intensity of joy feels quite normal and not outrageous.

b) Identifying the positive aspects of any person or situation and giving thanks for those aspects.

c) Welcoming failures as the lessons without which we would not learn.

d) Reframing difficulties as teachers in order to learn from them.

Choosing Your Feeling Expression

In order to make the Butterfly Shift effective as a mini-immersion experience of harmlessness, you will want to lead with emotion in order to warm up the interactive energy and capture the other person's attention. In choosing your positive emotion, there are three criteria that may help you:

1. Choose an emotion that is compatible with the type of Butterfly Shift you have chosen.

    The type of shift will depend on the person you have chosen to notice, as explored in Chapter Five.

2. Choose an emotional intensity that at least matches the intensity of the emotion of the person you are noticing.

    If your emotion is less intense, it will be easily discounted or ignored. If it is more intense, though, it provides a kind of wake-up call and captures the attention as long as it is not so intense that it is intimidating.

3. Choose a nuance that you can control.

    Since we are less experienced with positive emotions, you need to be sure that you can project the emotion you choose intensely and persistently.

Once you choose the emotion, it is important to sustain the feeling tone long enough to elicit a positive feeling in response from the other being noticed.

Ensuring the Success of Step Two

In a successful Butterfly Shift mini-immersion, what we notice or think about has to be colored by emotional energy.

The Institute of HeartMath has developed a number of techniques for practicing various emotions. One of the techniques most relevant to the Butterfly Shift is known as the Heart Lock-In, which can help you express a positive emotion consistently.[6]

Many of us have much less experience with positive emotions than with negative emotions. Practicing various positive emotions may feel fake or contrived at first because we are stretching an emotional muscle. But, just as with physical exercise, using that positive emotional muscle will lead to it feeling more natural in time.

Step Two reminds us that engaging emotionally strengthens our energetic connection with others, which is a critical component of an ethic of harmlessness. Emotion imbues our interactions with a different level of engagement and meaning. Our ability to express a wide range of positive emotions, instead of being energized mainly by negative ones, is crucial in experiencing harmlessness.

EIGHT

Reviewing Our Action Options

Once we have picked our focus and are aware of our chosen emotion, we need to select how we will execute the Butterfly Shift. Our action needs to be brief so that we are not seen as disruptive. It also needs to create a positive memory. And, when practiced over time, it needs to gradually strengthen our sense of the attractiveness of living harmlessly and the resolution required to do so.

Becoming Noticed

Noticing goes two ways. Not only do we need to pick a recipient, but the person we pick needs to notice us enough that our interaction is meaningful. So what will entice them to focus on us or to remember us? How will we ensure that the other person becomes aware of our intervention . . . in a positive manner?

People are most comfortable with, and likely to respond to, people like themselves. So we can become more "visible" to the recipient if we behave in ways that feel familiar. The most important first step is to hold an internal image of the other person as someone we already know and like as though they were a favorite family member. That automatically puts us at ease and begins to communicate warmth and caring.

We can complement this personalized connection with other behavior that helps create rapport, such as:

- o Maintaining appropriate eye contact

- Relaxing our muscles so that we are at ease
- Speaking clearly
- Limiting our feedback to 10 seconds or less without sounding rushed, which takes practice.

Exercise: Becoming Noticed

The next time you are in a group setting with people you don't know, try the following:

Stand quietly to the side, out of people's line of sight, and experience not being noticed.

Now purposely move towards someone to engage them in conversation. What initially captures their attention and makes them willing to talk with you?

Try the sequence a second time, and see if you can engage the other person more quickly and completely.

What does your experience tell you about how you can best become noticed by recipients of the Butterfly Shift?

Rapport is a critical part of building a memorable relationship. We've already discussed our tendency to group or "chunk" information. Rapport building requires us to "de-chunk" and recognize people as individuals ("the figure"), not just as part of a group background. Service quality research has shown that people re-evaluate their encounters with others afterwards based in part on the degree of rapport they felt.[1] We also know from research that if we create an actively positive impression, we are more likely to be remembered by the person we are targeting for the Butterfly Shift dynamic.[2]

Addressing a Person's Needs

Our actions will be most relevant if they are linked to the other person's needs and motivation. We can use the Maslow needs hierarchy—physiological needs, safety needs, belonging needs, status needs, and self-actualization needs—for this analysis.[3] The table below illustrates how we might detect each type of need and which type of shift might be indicated.

Identifying People's Needs

Person's Need	Person's Behavior or Appearance	Relevant Shift
Physiological	Looks haggard & worn out, unkempt, as though they are barely scraping by	Compassionate
Safety	Is careful to be accurate & not make mistakes	Compassionate or Grateful
Belonging	Is very sociable; interacts with customers & colleagues	Grateful
Status	Is dressed for success	Grateful or Joyous
Self-actualization	Is curious & asks questions	Joyous

In some situations, how employees dress or behave is strictly controlled by their employer and so it is not very informative from a motivational perspective. But, with careful focus, it may be possible for us to detect which needs or values are important to a person, even in a routine service situation. For example, you might see two airline staff checking passengers—one of whom has dark circles under her eyes and is carefully going through the ticketing without much eye contact (a candidate for the Compassionate Shift), and the other one who is joking with colleagues while quickly getting the passenger checked in and offering a better assigned seat (a candidate for the Grateful Shift).

Actions and Values

Actions are guided by our values. While there are many different value systems, there are two models that can help us move towards a possible set of universal values. One is embedded in the UN Universal Declaration of Human Rights, which is a model that has been endorsed by the 192 Member States of the United Nations. The other is the seven laws or values that comprise the Aboriginal sacred teachings. Below are examples from these two models of values that we might choose to guide our actions.

Examples of Possible Universal Values

Universal Declaration of Human Rights[4]	Aboriginal Sacred Teachings[5]
Dignity	Love
Worth of the individual	Respect
Safety	Courage
Equality	Honesty
Privacy	Wisdom
Freedom of thought	Humility
Freedom from harm	Truth

The more our actions are in keeping with these universal values, the more powerfully they will resonate with the recipient of the Butterfly Shift.

Cultural Filters

Who we notice, what we notice, and how we express respect are all related to our cultural values and conditioning. Our cultural socialization provides us with an often unconscious worldview through which we perceive ourselves and that shapes what we consider to be appropriate behavior. Cultural differences expose the fact that awareness is not the same as understanding and that behavior can be interpreted in many different ways. In fact, recent research has shown cultural differences in the range of facial cues that people use to read expressions of emotion.[6]

Our nonverbal behavior or paralanguage, which constitutes 60 percent of what we communicate, is directly influenced by culture. This easily-misinterpreted behavior includes how close we stand to another person, how much eye contact we maintain, whether and when we touch others, how we greet others, and many other nuances. In communities with large immigrant populations, cultural sensitivity to relevant nonverbal cues is especially important.

Cultural values act as filters in terms of how we interpret what we see in interpersonal interactions and what we think is happening. In turn, cultural values form a filter through which our actions are interpreted by others. While different cultural values may often seem irreconcilable, especially between Western and Eastern cultures, they actually represent a difference in emphasis.

Our current understanding of the role of cultural values has its roots in the early research of anthropologists Edward

Hall[7] and Florence Kluckhohn with Fred Strodtbeck[8] who helped describe the wide range of value structures and their common underpinnings. Social psychologist Geert Hofstede[9] was one of the first to study cultural value variations specifically within a work context. Combining the work of these experts, we can identify five key values continua along which behavioral expectations differ among cultures in ways that are relevant to the Butterfly Shift dynamic: (a) the degree of role hierarchy or power distance; (b) individual versus group needs; (c) tradition versus change; (d) efficiency versus effectiveness; and (e) assignment of responsibility.

The Degree of Role Hierarchy or Power Distance

When we interact with others, we make assumptions about how much deference they should display towards us and why. One cultural extreme is the strictly hierarchical "chain of command," while the other extreme is a completely egalitarian approach. Reflected in this values continuum is a society's attitude towards "face" or *miànzi*—that is, the social perception of one's prestige or authority. While *miànzi* is specific to Chinese culture, similar concepts exist in Indian, Japanese, Korean, Thai, Vietnamese, and Middle Eastern cultures. The closest equivalent in the more egalitarian cultures of Europe and North America is that of embarrassment, but embarrassment is more of a personal emotion and does not carry with it the same public loss of authority that is implied by *miànzi*. There is also the concept of the loss of credibility, but that loss in western cultures is based on actions by the individual that demonstrate inconsistency between word and deed rather than on a more general undermining of social status.

In high power distance societies like China or Japan, we would expect all service providers to act with extreme deference (being lower in the hierarchy than those they serve). So we would evaluate whether or not someone was being unusually helpful through a filter that assumed extreme deference as a normal part of service.

In low power distance societies, though, we might interpret such deference as coldness. Unless we were in an extremely formal setting, we would be expecting more of an egalitarian, chummy exchange. In this instance, we might dismiss the person as unreceptive to the Butterfly Shift interaction when, in fact, they could be a good recipient.

In order to execute the Butterfly Shift dynamic, we need to be careful to draw correct conclusions about the degree of extraordinary helpfulness we are being offered. In turn, we need to be careful how we communicate respect to the recipient and how our enthusiasm or gratitude would be received. If we are unsure of the cultural context, being both formal and warm usually ensures that our actions are seen as respectful.

Individual Versus Group Needs

In cultures where individualism is valued, we would credit the individual with any special initiative. Being remembered individually by staff at establishments that we frequent would be considered unexpected and an extra level of assistance.

In cultures where individualism is *not* valued, persons are viewed as extensions of their group, whether an organization or family. Boundaries between "in-group" (defined usually as the extended family) and "out-group" are well defined, and the interests of the "in-group" are placed ahead of those not in the "in-group." If we were regular customers (and so, "in-group") of an establishment, we would expect our special status to be acknowledged without having to make a request. We might expect to be greeted or acknowledged by name or to be given special consideration (as happens with gold- or platinum-tier frequent flyers). This would be considered usual service, not exceptional.

In terms of the Butterfly Shift dynamic, we need to be careful what weight we assign to any special acknowledgement we receive. By contrast, we might have a particularly positive impact on an "out-group" recipient who is not expecting us to notice them—for example, calling them by name.

Tradition Versus Change

Members of traditional or indigenous cultures tend to value a high degree of predictability and continuity, a long-term perspective, and a minimum of risk. As a corollary, the wisdom of elders is revered, and the impact of new initiatives is carefully considered into the future (often noted as being for seven generations).

Members of cultures that value risk tend to glorify the energy of youth and to value "new is better." In such cultures, elders are usually dismissed as being old and "out of date."

In terms of the Butterfly Shift dynamic, we need to be conscious of age differences and be particularly attentive to elders in contexts where they are viewed as the ultimate authority. If we find ourselves in a situation where the other person's behavior seems excessively rule-bound, we can reframe the situation to express appreciation for attention to consistency and to solutions that are sustainable over time.

Efficiency Versus Effectiveness

"Being" cultures value the quality of interaction rather than what is accomplished in the interaction. Feedback is usually provided privately in order not to undermine the relationship or cause the other to lose face.

"Doing" cultures, on the other hand, value successful task outcomes even when achieved at the cost of undermining a relationship. Staff members are encouraged to "speak their minds" and to "be honest." Feedback is often provided publicly, characterized as being objective; and the recipient is expected to receive it as input for how to increase efficiency.

There are two aspects of the Butterfly Shift dynamic that are influenced by this cultural value continuum. First, the focus for our feedback needs to include the impact of both the process or quality of the interaction ("being") and the outcome or efficiency of that interaction ("doing") in order to cover the whole cultural continuum. Second, we need to be sensitive about the context in which feedback is given. In a culturally ambiguous setting, we may be most successful if we provide feedback to the person in a relatively private setting even though the feedback is positive.

Assignment of Responsibility

In "high-context" cultures, the meaning of a communication is inferred from its context. These cultures are group-oriented, and interpersonal networks are of particular importance. Superiors have responsibility for the actions of their subordinates, in the same way that we hold parents responsible for the actions of their children. We see repeated examples in Asian news sources where corporate or political leaders state publicly that they are accountable for the wrongdoings of those under their supervision. To shirk accepting this responsibility would be considered unethical.

In "low-context" cultures, everything is stated explicitly, and task efficiency is valued more highly than interpersonal relationships. Staff are assumed to be individually responsible for their actions despite any supervisory negligence. In such cultures where context is largely ignored, any wrongdoing is researched until the responsibility can be fixed on a specific culprit—typically the most junior staff member involved.

For purposes of the Butterfly Shift, it is important that we attribute positive responsibility to the appropriate person and that we recognize when a person has extended themselves to meet the spirit, not just the letter, of the situation.

> ### Exercise: Cultural Flexibility
>
> Pick one of the following value dimensions and practice interacting in a manner that is the opposite of your cultural conditioning:
> - Degree of role hierarchy
> - Individual versus group needs
> - Tradition versus change
> - Efficiency versus effectiveness
> - Assignment of responsibility
>
> From your experience with this exercise, what could you do to increase your ability to adapt your behavior to different cultural settings?

Feedback and Harmlessness

The Butterfly Shift dynamic—as a mirror of a milieu of harmlessness—is not only about our own actions but is also about enabling others to act. Unless we know what makes a positive difference in others' lives, we cannot intentionally replicate what we have done. The same is true for others. Providing positive feedback is an important way in which we enable others.

Feedback is the process of communicating to others the effect of their behavior on us. The point of feedback in the Butterfly Shift dynamic is to engage both the person's desire to be helpful and the desire for recognition because those two dynamics are what will result in a repetition of the desired behavior. There are seven characteristics of effective feedback that are relevant in the Butterfly Shift:

1. *Immediate*

    Feedback needs to be provided as soon after the interaction as possible. If we wait too long, it is difficult for the person to link the feedback to what they actually did.

2. *Sincere*

    Feedback needs to reflect how we actually feel or it will sound phony or contrived. This is particularly important in the *Compassionate Shift*.

3. *Consistent*

    Feedback needs to be given each time the person is appropriately helpful. If we provide positive feedback one day and then are silent the next day when the person is precisely as helpful, we are communicating that the helpfulness didn't really matter. This characteristic becomes particularly important in the *Grateful Shift*.

4. *Unconditional*

    Feedback needs to be given whether or not the other person responds positively. In other words, we don't pause part way through to see if the other person is going to be grateful to us. This characteristic is particularly important in the *Joyous Shift*.

5. *Relevant*

   Feedback needs to be about a behavior that the person controls so that they do have the option of repeating it.

Exercise: Giving Feedback

Pick a recent time when you gave feedback to another person and evaluate that feedback. Was it:

- Immediate?
- Sincere?
- Consistent?
- Unconditional?
- Relevant?
- Appropriate?
- Specific?

If you answered "no" to any of the characteristics, what could you do to improve the feedback you give?

---

6. *Appropriate*

   Feedback needs to match the emotional intensity of the feedback with the level of effort shown by the person.

7. *Specific*

>Feedback needs to identify precisely and accurately what it was that made a difference to you, rather than being general ("that was great").

Actions and the Joyous Shift

The action step of the Joyous Shift is a bit different from that of the other two shifts because here we ourselves are taking action, not just providing feedback on the recipient's action. The purpose of our action is twofold. *First*, we are giving the recipient a clear example that the underlying dynamic of the world is abundance rather than scarcity. And, in so doing, we are providing ourselves with a reminder regarding the basic dynamics of harmlessness.

Second, we are providing a chance for the recipient to experience joy—an unexpected, overflowing feeling of happiness. When we do so, we ourselves also experience that overflowing joy.

In many ways, this component of the Joyous Shift is similar to the performance of random acts of kindness.[10] We are the ones initiating the kind act, the unexpected assistance. What makes it different is the lack of anonymity. It is wonderful to surround others with unexpected kindness. What we are attempting in the Butterfly Shift, though, is to help people become conscious of their ability to provide that joy for others. So we need to accompany the action with a prompt like, "If this has delighted you, do the same for someone else." Thus, we are creating a complex milieu of harmlessness for ourselves and others.

What we wish to do in the Joyous Shift is more like the "pay it forward" initiative that began with Catherine Ryan Hyde's novel, *Pay It Forward*. The premise in the novel is that you can get exponential benefits if you do a favor for someone with no expectation of being paid back but, instead, ask that person to do a favor for a different person (ideally three different people). There is now a Pay It Forward Foundation that supports "pay it forward" initiatives by school children, modeled on the movie made from the novel.[11]

> ### Exercise: Initiating Change
>
> During the next week, pick a situation where you can perform an anonymous kindness. Observe what the other person does as a result. Do they notice? Do they act in a more kindly manner?
>
> Pick another time when you can perform an unexpected kindness and ask the person to "pay it forward." Observe how they respond. What is the likelihood that they will act in a more kindly manner?
>
> What conclusions can you draw from the two situations?

Detachment and the Butterfly Shift

We've already mentioned the importance of not requiring that the recipient be grateful. Keep in mind that the recipient has not asked to be chosen. That person is just going about their everyday business. We are the ones who have decided to execute a Butterfly Shift that involves them.

If we have chosen well and expressed the accompanying positive emotion, there is a strong possibility that the other person will be pleased and respond. But they may be too preoccupied. We cannot make their response a requirement. We need to leave them room to freely choose not to participate without any negative consequences.

> **Exercise: Detachment**
>
> Think back over the past 48 hours and pick a time when you did something for someone else with no thought of what you might get in return and no vested interest in their response.
>
> Now think about a time when you did something for someone else in order to get them to act in a particular way.
>
> What was different for you in the two situations?
>
> What helped you remain detached in the first scenario?

The dynamic we are referring to is known as detachment. Detachment is not the same as indifference. Detachment is a caring response, but it is one of respect. We surround the person with the potential for a positive experience while remaining unattached to the actual outcome.

Facilitating the Mini-Immersion

Choosing our action is the final step in generating the mini-immersion of the Butterfly Shift. We need to make sure that the person who is the focus of our attention is aware of us. Our timing is critical, as is being warm and engaging and culturally relevant. Our overall intention is to create an engaging experience of what harmlessness feels like—its positiveness, its feeling of openness and well-being.

What differentiates the Butterfly Shift from routine positive encounters is that we purposely provide specific feedback on why we are doing what we are. It is this deliberateness that brings with it the potential to initiate an ever-expanding ripple of goodwill, expanding into an ethic of harmlessness.

NINE

Acting — Step Three

Actions speak louder than words. We may feel compassion or gratitude or joy but, unless we actually do something as a result, what difference does it make?

Through the first two steps of the Butterfly Shift, we have become mindful of the small positive contributions of those around us and have engaged our own intense positive emotional response. Now Step Three is our actual outward execution of the Butterfly Shift so that it is visible to others.

Ways to Improve Your Action Effectiveness

Step Three has several inherent challenges: engaging the person we are targeting with the Butterfly Shift, conveying what we wish to communicate, and acting in a congruent manner. In addition to the strategies we have already discussed for the first two steps, the following are ways that we can enhance the successfulness of our actions:

1. *Practice giving useful and relevant feedback.*

    Since we have more experience with negative criticism or hearing what we have done wrong, feedback on what we have done well is particularly important. The critical components for us to practice are:

    - Setting the stage so that the recipient knows that what is about to be said is positive, not negative: "I want to thank you. . ." or "I want to compliment you . . ."or "I can see you are in a difficult situation and I want to . . ."

- o Stating clearly what we liked about the encounter and how it made us feel, the positive difference it made for us.

- o Being specific instead of just saying, "Thanks," or even "Thanks, you were helpful." Get in the habit of detailing *how* they were helpful.

2. *Focus on behavior that could be repeated*

While it is nice to congratulate someone on unusually helpful behavior, what we want to identify is behavior that the recipient could replicate over and over again.

3. *Adapt your approach to the person's apparent needs.*

Here are some examples of ways that you could respond, which link to the person's apparent need:

Responding to People's Needs

Person's Need	Your Response
Physiological	Focus on their well-being & putting them at ease: "You look really tired."
Safety	Act in a manner that is non-threatening: "I just want to let you know how much I appreciated your time in helping me find what I wanted."
Belonging	Interact warmly: "You really made my day by solving this billing issue."
Status	Let them know that you value their opinion: "Your suggestion last week that I buy the dual handset was excellent. It is working out very well."

4. *Experiment with different cultural factors.*

    Depending on your own cultural background, select less familiar cultural response patterns to practice. There are two common patterns that you could try:
    - Respond to both the interpersonal ("being") and the task efficiency ("doing") aspects: "Thank you for being so welcoming and also for the quick service."
    - Remain respectfully formal while also being warm.

5. *Remain alert for teachable moments.*

    Look for times when the other person's extraordinary helpfulness is exceptionally clear and they are available for feedback.

6. *Focus on creating an enabling environment.*

    We cannot force another to change, and that's not actually the point. The Butterfly Shift is as much about changing ourselves—building the habit of harmlessness—as it is about creating the possibility for change in others. We do enable others, though, by focusing on small, easily replicable actions that make a significant positive difference.

7. *Create a personal connection.*

    Many of our daily interactions with others take place in an impersonal environment. We can change that by interacting with the recipient by name.

    If you don't know the name of the person who is being helpful, you can ask, "I'm sorry, but I don't know your name." When they offer their name, you can offer yours in return, matching the pattern—"Jane" for

"Joe," or "Jane Nelson" for "Joe Green"—and then thank them by name.

Action and the Type of Shift

Since the purposes of the three types of shift are different, our actions will also be different. Here are some examples of different strategies:

1. For the *Compassionate Shift*:
    - Begin with the difficulties you have noticed: "That customer was certainly rude. I don't know how you remained so pleasant."
    - Express specific appreciation for the extra effort the recipient has made: "Thanks for getting me an invoice even though your computers aren't working."
    - Provide feedback on the difference it made to you: "Your willingness to take the time to separate the transactions will really help me with my accounting records."

2. For the *Grateful Shift*:
    - Begin with the extra effort you noticed that the recipient made: "Thanks for the free car wash. I know you didn't have to do that when all I came in for was an oil change."
    - Express your gratitude for it. "I'm really grateful for your thoughtfulness."
    - Provide feedback on why it made a difference: "I feel like I'm ready to drop I'm so tired. Your carrying out my packages was just the helping hand I needed."

3. For the *Joyous Shift*:
    - Notice and act on the potential to be especially helpful — pay the balance of the recipient's bill when they are struggling to find the change, carry the parcels of a person who is juggling a number of them, take a hot meal to a neighbor whose power is out.
    - Acknowledge that you did it purposely and why: "We got our power back a couple of hours ago, but I noticed yours was still out and thought you could use some hot food."
    - When thanked, prompt the other to pay it forward: "If you really want to thank me, do a similar favor for three other people."

Choosing the Type of Shift

In choosing the type of shift, we need to keep in mind what will allow use to create mini-immersion experiences in harmlessness for ourselves and others. Here are some suggestions about making a shift selection:

1. For the *Compassionate Shift*:
   Choose this type of shift when you have some extra energy to draw on to focus on the interaction.

2. For the *Grateful Shift*:
   Choose this type of shift when you can sense your own gratitude through internal clues such as relief or relaxation of tense muscles or profound pleasure for the help given.

3. For the *Joyous Shift*:
   Choose this type of shift when you have the opportunity to be unexpectedly helpful *and* either you are

interacting with the recipient or you can provide the nudge to pay it forward through the use of a tool like the Smile cards[1] (which ask the person to pay it forward).

To maximize your ability to create an experience of overall harmlessness, you will want to rotate the type of shift you choose so that you practice each type at least once each week.

Ensuring the Success of Step Three

A successful Butterfly Shift dynamic requires us to notice our context in a flexible manner and with a positive emotional state. We are most likely to be successful if we concentrate on engaging respectfully with those ever-present helpers who are usually taken for granted and treated as invisible. Cultural filters become important both in how we interpret what we observe and in how we respond.

What we are doing in this daily practice is not only to create an alternate experience of harmlessness but also to increase its frequency and intensity so that it becomes our default reality. Each day that we practice the Butterfly Shift, we help create a positive harmlessness immersion experience for ourselves and others. The key is that we do this without any expectation of thanks. We actually do it primarily for ourselves—to shift our base state from harmful to harmless. Through the Butterfly Shift, we create the possibility of more positive action in the future. We continue to fill up the pail with clear harmless energy.

Part Three

Ensuring Ongoing Harmlessness

TEN

Maturing Into Harmlessness

Creating harmlessness mini-immersions via the Butterfly Shift is the first step in changing our awareness. But such brief experiences are not enough to anchor harmlessness as our societal norm. To do that, we need to transform our shared worldview from one where harm and violence are assumed to be inevitable to one that affirms harmlessness as our chosen ethic.

Our worldview is a framework of attitudes, beliefs, values, and presuppositions through which we interpret our experience and choose how we interact with others. It provides us with answers to basic questions like, "Why am I here?" or "Where am I going?" or "How should I attain my goals?"

Part of our worldview is our assumption about what maturity means and what we are developing towards. Given the Principle of Nonlinearity, it is important that we find a way to conceptualize psychological maturity that does not assume a linear sequence of developmental steps. In its Values and Lifestyle (VALS) Program report, the Stanford Research Institute offers the following definition of psychological maturity, which can provide us with a starting point: "A progression from partial toward full realization of one's potential . . . [that] involves a steady widening of perspectives and concerns and a steady deepening of the inner reference points consulted in making important decisions. Thus, the role of habit and 'stock answers' abates as a person matures, and the person becomes increasingly more complex and self-expressive in a values sense. . . . The more developed a person is, the more complex his or her value structure and the more diverse the range of value-based reactions."[1]

This definition of psychological maturity is useful because it does not assume a particular sequence of developmental stages. It also provides us with some guidelines for redefining our developmental process as one of increasing complexity. The goal proposed is that we realize our individual potential, not reach some generalized standard of adulthood. In order to do so, we need to expand the perspectives from which we can view life and understand issues, thus becoming more flexible and adaptable in our approach. At the same time, we need to strengthen our inner values matrix and sense of social responsibility so that we are able to identify and act on the nuances presented.

If we want to change people's worldview so that harm is no longer acceptable, then we have to change our shared definition of psychological maturity so that it supports increasing complexity. We need to re-examine our presuppositions about the focus and context of maturation, where we are starting from, our relationship with others, and our developmental process in light of what we know about the principles of our cosmos.

The Focus of Our Developmental Models

Not surprisingly, our view of psychological maturity has been shaped over the past 200 years by assumptions from Newtonian physics. The goal we have been taught is that of the "rational man" who grows through a series of predetermined, hierarchical stages to become objective and independent by the time he—or she—is of legal age. Moving beyond this limited view of ourselves is challenging, but several critical shifts in psychological theory have helped lay the groundwork.

The assumption that we are fully developed by the end of adolescence has been imputed to a range of early well-known psychological theorists—for example, Freud, Piaget, and Binet. The *first* shift that occurred in psychological theorizing was moving from an exclusive focus on child development to an interest in ongoing adult development as well.

The psychologist Erik Erikson was one of the first to propose stages of development in adulthood, and he held the then radical belief that cultural and societal influences play a role in shaping us.[2] Daniel Levinson, one of the founders of the field of positive adult development, expanded on Erikson's work and proposed a number of "seasons of life"— preadulthood, early adulthood, middle adulthood, and late adulthood.[3] Psychologist Bernice Neugarten went further by shattering stereotypes of those over 55 as being elderly, feeble, and expendable and coining the term "young-old" (similar to the term "active seniors") to describe those who lead vigorous, self-fulfilled lives of service to their communities.

Work in positive adult development then expanded to include a professional association and a professional journal.[4] By the late 1990s, the focus on adult development had broadened further to include positive psychology.[5]

We now see an increasing emphasis on lifelong learning— an ongoing growth process throughout our lives to support our realizing our full potential. Sweden's study circles—small groups of people who meet repeatedly to explore issues of common interest—is one model of lifelong learning structures. In its 2006 report, *Adult Learning: It Is Never Too Late*, the European Union encouraged additional initiatives to stimulate ongoing learning. Much of the focus has been on adults over 55 years of age, with a range of learning settings such as the Elderhostel Institutes, the Institutes for Learning in Retirement, and the Universities of the Third Age. Institutions such as the University of Delaware and the University of Toronto now have an Academy of Lifelong Learning.

The *second* shift has been from a preoccupation with pathology and the resolution of conflicts to an examination of happiness and people's desire for self-actualization. This shift can also be characterized as refocusing from weaknesses to strengths, from deficit to abundance.

> **Exercise: Assumptions About Development**
>
> When you think about your own maturation process:
>
> 1. How long do you expect to keep maturing?
>
> 2. Do you expect yourself to mature primarily? By developing strengths? By resolving conflicts? Both?
>
> 3. Do you assume psychological maturation includes emotional, mental, and spiritual maturation? Why or why not?

Abraham Maslow, whose hierarchy of human needs we have already reviewed, was among the first to focus on human potential and the possibility of peak experiences.[6] He is credited with founding humanistic psychology in the 1950s as a "third force," moving beyond Freudian theories and behaviorism. Transpersonal psychology has since developed as a "fourth force," moving beyond humanistic psychology's emphasis on self-actualization to include the soul. It focuses on "the study of humanity's highest potential, and . . . the recognition, understanding, and realization of unitive [tending to promote unity], spiritual, and transcendent states of consciousness."[7]

Our Developmental Context

Our main psychological theories are silent on whether or not this physical life is the sum total of our experience. But recent research in the empirical sciences suggests that physical reality and spiritual reality are closely intertwined and that continuity of consciousness over multiple lifetimes—or reincarnation—may be more than a metaphysical hypothesis.[8]

From a metaphysical perspective, this life is generally viewed as only one of many in which we learn and contribute. Our engagement in a number of lives allows us to learn to be both unique (our own person) and part of the whole cosmic energy field (the Principle of Interconnectivity). For those who ascribe to the theory of continuity of consciousness, it is the soul that provides the link from life to life, carrying within it a memory of life lessons learned and the chosen life purpose. We do not necessarily remember these other lives; indeed, it could be confusing if we were constantly having to sort out current from non-current happenings. Rather, our focus is expected to be on the karmic lessons that we, as souls, have set for ourselves in this lifetime.

In thinking about continuity of consciousness, it is easy to slip into the illusion of linearity. We talk about "past lives" or "past life" readings. But the Principle of Nonlinearity reminds us that our sense of linear time—past, present, future—is only an organizing principle for our convenience, not an objective reality. In actuality, everything is happening at once. The author and psychic Jane Roberts provides us with an intriguing view of this simultaneity in *The Education of Oversoul Seven*, which portrays the struggles of a discarnate entity

(Seven) responsible for four human "incarnations" living simultaneously in time periods ranging from 35,000 B.C. to the twenty-third century A.D.

How we conceptualize our context may influence our approach to ethics. If we accept the idea of continuity of consciousness, we are then presented with the companion concept of karma,[9] or the effects of our choices. We become aware that sooner or later—in this life or another one—we *will* experience the consequences of our actions . . . for better or worse.

Exercise: Developmental Context

What is your assumption about continuity of consciousness—do you believe this is your only life or simply one of many?

What implications, if any, does your assumption have for how important you feel it is to behave harmlessly?

Revenge, or returning harm for harm, can feel more or less important depending on our assumptions about context. If we or persons we care about are badly harmed, revenge may seem justified in order to make sure that the perpetrators pay for what they have done, even though we have developed social institutions to provide justice. If, however, we believe the proverb "what goes around, comes around," we will anticipate that karmically the perpetrators will reap the consequences of their choices and this may make it easier not to seek revenge. Being able to release the desire for revenge is critical since attachment to revenge actually harms us by keeping us fixated in the past.[10]

Another consequence of a belief in continuity of consciousness is that it lays the groundwork for compassion and forgiveness. We truly don't know where another person is in their journey. Perhaps they are a "young soul," just learning to manage their own physical existence. Perhaps they are an "old soul" who should already know better but who needs another lesson. Whatever the reason, we can recognize that we are each in the process of developing and will each make mistakes.

Our Starting Point

Most psychological theories assume that life begins at our physical birth, which makes sense if we focus only on the existence of this physical body. Michael Cremo has phrased an alternate proposition in his book *Human Devolution*, based on Vedic teachings: "We do not evolve up from matter; rather we devolve, or come down, from the level of pure consciousness. Originally, we are pure units of consciousness existing in harmonious connection with the supreme conscious being."[11] Instead of assuming that our starting point is our physical formation as a physical embryo, we could describe our "birth" as the coming together at a point in time of mind (cosmic mental energy), soul (our continuity of consciousness), and a physical form.

Of the psychological theorists who include the soul as part of who we are as humans, few provide a framework that assumes development prior to physical birth. Rather, they start at birth and then trace a strengthening connection with our spiritual self. One exception is the philosopher Michael Washburn who posits a Dynamic Ground (similar to cosmologist Brian Swimme's "all-nourishing abyss"[12]) in which we are embedded prior to birth and from which we emerge at birth.[13]

Transpersonal psychology has introduced the mind-body-spirit model. If we think of "spiritual" as referring to our connection to the cosmic energy field (the Principle of Interconnectivity), we can discuss this dimension without becoming mired in the conflicting philosophies of the world's religions.

One of the first psychologists to propose a theory that included the spiritual component of our lives was Roberto Assagioli.[14] In his book *Psychosynthesis*, he proposed both a personality and a soul, providing a model of the person that has become known as the "egg diagram."[15] This diagram positions the essence of a person, comprised of consciousness and will, within the influence of the Higher Self or the soul. Assagioli's focus was on self-realization, which includes spiritual development beyond self-actualization.

Since Assagioli, other theorists have challenged our understanding of the boundaries of our being. For example, Ken Wilber, the author of Integral Theory, has proposed the AQAL (All Quadrants, All Levels) map of human experience in which his final stage of consciousness is the transcendent/soul/unitive view. The nine stages of ego development proposed by the educational psychologist Susanne Cook-Greuter stem from Wilber's model, with her ninth stage being unitive transpersonal development.

Management consultant Richard Barrett has stated, "The soul is our true self—our essential nature—a vortex of energy manifested in physical form. The transformation from ego-based consciousness to soul-based consciousness occurs when the ego is able to transcend its fear-based existence and begins to look for a deeper meaning to life."[16] So we have another contributor proposing that we are comprised of not only a physical body but also a spiritual component.

> **Exercise: Our Starting Point**
>
> If you assume that you have lived more than one life:
>
> How would you identify what psychological strengths you have already developed in other lives?
>
>
> What role do you expect the soul to play in your psychological development?

What is not clear in these various psychological theories is any method for understanding how learnings from other lives might lay a foundation for our current life. In that regard, we remain dependent on metaphysical theories. In the Ageless Wisdom tradition, the cosmic energy field is said to be comprised of seven types of energy streams, or Rays.[17] Individual souls are said to belong to one of the seven ray groups of soul energies, and are further described as having a personality ray (related to the developmental tasks of this lifetime), supported by rays of the mental, emotional, and physical bodies. This ray structure not only engages learnings from other lives but provides a range of possible personality configurations and life focuses. Esoteric astrology complements this approach by proposing that one's birth chart provides information on lessons already learned and those posed for this lifetime, with the Ascendant or Rising Sign[18] representing soul purpose in this life.

Our starting point becomes important because it determines whether or not we must assume that we all start with a "blank slate" (as the philosopher John Locke called it) or whether we can assume that most of us enter this life with a developmental background in place from other lifetimes. If the latter is the case, then we can be more ambitious in how we embed harmlessness in our social code.

Maturing in Relation to Others

Fundamental to most psychological models is the belief that the ultimate goal of adult development is to construct a separate, independent self-identity with clear boundaries between self and the "objective" world. But is such a belief in keeping with the seven principles (discussed in *Principles of Abundance for the Cosmic Citizen*) that underlie the functioning of our cosmos?

The Principle of Interconnectivity tells us that we are in fact all part of the same cosmic energy field. Therefore, our maturational tasks include being able to manage the tension of uniqueness and being part of a larger whole. When we ignore this basic interconnection, we create an illusion of separativeness that is counterproductive for us.

The Principle of Interdependence tells us that we are interdependent, not independent. But we face a challenge because we are not necessarily conscious of that interdependence. We assume that we *should be* separate, independent entities, which feeds the "us-them" mentality that underlies violence. But our cosmos is constructed such that we are *already* interdependent. That is not something that we have to mature towards. However, experiencing our interdependence is not easy. Interdependence requires that we ensure that all parties benefit, not just ourselves. It means that we accept that we are mutually responsible emotionally, physically, economically, and morally.

The Principle of Participation tells us that there is actually no "objective" reality and that we create our perception of reality through our observation and focus. Why, then, would we wish to establish clear boundaries between ourselves and the all-nourishing abyss through which we are connected with all life? A more useful relationship model might be the semi-permeable membrane of our cells, which lets through what is beneficial and screens out that which isn't.

> **Exercise: Relating to Others**
>
> Assuming that we are interdependent, list three implications for how you would interact with others harmlessly rather than harmfully:
>
> 1.
>
> 2.
>
> 3.

In the context of maturing towards harmlessness, there is also the issue of how we define "others." In general, psychological theories are preoccupied with how we relate to other humans. But not only are we linked energetically to all life forms, but we also know that we are not the only intelligent species on Earth. Emory University neuroscientist Lori Marino has raised ethical concerns about the practice of capturing and confining dolphins to perform for humans. "Dolphins are sophisticated, self-aware, highly intelligent beings with individual personalities, autonomy and an inner life. They are vulnerable to tremendous suffering and psychological trauma."[19] Such data suggest that we may also need to consider an interspecies ethic of harmlessness.

Although we talk more about interdependence, it has not yet influenced the stated underlying goal of psychological development as presented in psychological theories—either in relation to humans or to other intelligent species. By and large, we still see the process of meeting our own needs as separate from our responsibilities to others . . . except in a family context. One of the exceptions, however, is the growing group of persons called Cultural Creatives.[20] These are people from all walks of life who demonstrate both a commitment to a spiritual practice and an active concern about social issues.

How we understand our relationship with others is central to our ability to act harmlessly. As long as we see ourselves as separate and independent entities, we will feel free to maximize our self-interests at the expense of others.

Our Developmental Process

The nature of theories is to look for common characteristics or experiences from which we can describe a universal model. In the case of psychological development, this has resulted in theories that assume set stages that we all go through in sequential fashion. There is a certain degree of determinism involved since the successful completion of one stage is required by, and built on, in the next stage—that is, the stages are "nested."

But the Principle of Nonlinearity tells us that our development is actually not linear, though it may appear linear from certain perspectives. So why would we subscribe to a psychological theory in which development is said to occur in linear, hierarchical stages?

Ken Wilber, in proposing levels in his spectrum of consciousness, argues that a hierarchical progression is inevitable. He takes examples from nature to illustrate that larger units are comprised of smaller units in a nested fashion.

By contrast, Richard Barrett has proposed a seven-stage developmental process, building on the needs hierarchy of

Maslow... but with a twist. The first three stages encompass the physiological, safety, belonging, and self-esteem needs of Maslow's model. At the fourth stage, he suggests a figure-ground reversal—a shifting of perspective from the personality to the soul, creating a disjunction in the development pattern. The last three stages represent the unfolding of the soul: internal cohesion, making a difference, and service.

In Barrett's Seven Levels of Consciousness Model, we see one of the first proposals (outside of the Ageless Wisdom literature) that growing as a personality concerned with the physical body's reality is only a small part of our developmental process. Instead, he focuses a good portion of his attention on spiritual development, positioning our maturation in a context of service rather than self-interest.

What is apparent from all the models of hierarchical stages is that we need new images for thinking about our development. While theorists may assume that we must master a set of skills at one level before moving to another level, in real life our mastery coexists at different levels for different skill sets. In a sport like figure skating, for example, this is acknowledged in scoring athletes separately on technical precision and artistic performance.

Perhaps we could think of our developmental process as more like a Chinese restaurant menu where one can select a specified number of item from each of several groups. We have categories within which we need skills mastery, but there are multiple options of how we can go about achieving that mastery.

When we really think about our growth process, it becomes obvious that we do not follow a simple sequential, hierarchical process. In order to change, we have to "come apart" to a certain extent so that we can reconfigure ourselves differently. Seldom can we simply build on what came before.

Sometimes the disruption, differentiation, or dissolution is forced on us, and sometimes we choose it.

Change can take many forms, not just one. If we take a musical analogy, we can change a piece of music by changing the pitch or the tempo, the volume or the rhythm.

> **Exercise: Developmental Process**
>
> Think of someone you consider to be your peer and think of at least three ways in which that peer is more skilled than you are:
>
> 1.
>
> 2.
>
> 3.
>
> Now think of at least three ways in which you are more skilled than that peer:
>
> 1.
>
> 2.
>
> 3.
>
> What do the above suggest to you about how we develop?

And there is not necessarily one right way to change or develop. While it may be true that there are basic building blocks that we need to have in place, they can be combined in a number of ways. Take, for example, the following words as building blocks: "con" and "tract" and "or." Each of these is meaningful in and of itself. They are also meaningful in various combinations: "contract" and "tractor" and "contractor."

There are also multiple models for what skills we need. Turning again to a musical analogy, a violinist has certain basic skills to acquire. But there are different methods and sequencing for learning those basic skills. Further development depends on what the violinist is interested in. For social events and personal pleasure, basic skills may be enough. To perform as a soloist, there are both technical and performance skills that will be needed . . . which are not necessarily the same as those needed to play in an orchestra.

One of our maturational learnings is that we don't have to be good at everything. Rather, we can focus on certain areas of competence and look to others to supply the skills that we lack. We see this in competitive racing where competitors will "draft" off of each other in order to maintain speed with minimal effort.

Another example may help us view the matter of sequencing differently. Take learning to drive a car. To drive competently, we need to be able to manage the amount of pressure on gas and brake pedals, monitor the rearview mirrors while also watching the road ahead, and develop the hand-eye coordination necessary for steering and backing up. As long as we can practice in a large safe area, does it really matter in what order we choose to master these skills? Of course, there are specific situations in which sequence matters. To drive safely, we need to be able to read the road signs so reading becomes a prerequisite. It may be that sequencing is important in early life when competence is linked to physical maturation, but not so important later in life as we focus on particular skills.

The level of skill that we need to achieve also varies. For everyday living, for example, being able to use a hammer and a screwdriver may be enough. But someone working as a finishing carpenter will need a wider and more sophisticated set of woodworking skills.

Finally, there is the issue that developmental skills are not necessarily additive. Sometimes the whole is greater than the sum of the parts. Take, for example, the making of basic cakes, biscuits, or pancakes. For the cake, we will need flour, salt, sugar, baking powder, butter, milk, and eggs. Subtract the eggs, reduce the amount of sugar and milk, and we have biscuits. Put the eggs back in but reduce the amount of sugar and baking powder, and we have pancakes. So, the same basic ingredients but a different outcome.

We find another example in choral music. Each of the parts—soprano, alto, tenor, bass—sound beautiful on their own. If they are joined in unison singing, we get a much richer sound than any part alone. But if the various parts sing harmony instead of unison, suddenly we have not only a richer but a qualitatively different sound.

We have already seen that the transpersonal perspective introduces the element of spiritual as well as physical development. In order to understand our maturational process from a spiritual perspective without getting caught in an illusion of linearity, we need to be able to describe multiple options in a non-theist manner. The Spiritual Evolution Assessment Scale was designed to provide just such a measure.[21] What it verifies is a multidimensional maturation model that fits well with the seven cosmic principles explored in *Principles of Abundance for the Cosmic Citizen*.

What difference does it make how we view our developmental process? As long as we view our developmental process as one of sequential stages, we are likely to judge one stage as better than, or more advanced, than another. We will begin evaluating and judging where another is in their developmental process, criticizing them for lagging or lauding them for leading. Such judgments are inappropriate and can be harmful. If instead we accept that we are each engaged in maturing in our own way and at our own pace, then we are more likely to stay focused on our own growth objectives and be supportive of others.

Harmlessness and Moral Development

Our moral development is of particular relevance to any discussion of harmlessness. The psychologist Lawrence Kohlberg is credited with opening the debate on how moral development takes place.[22] His stage theory assumes that moral behavior develops over time and is primarily focused on the mental determinants of what constitutes "justice." It assumes that children begin life as basically amoral and act initially to avoid punishment and that a recognition of interdependence does not emerge until one of the later life stages. And it is very compatible with the Darwinian perspective that we begin life as competitive, selfish individuals.

Carol Gilligan and Nel Noddings have proposed instead an "ethics of care."[23] This perspective is grounded in a recognition of interdependence as well as a belief that those particularly vulnerable to the choices made should receive extra consideration. The focus is on the relationship between the "one-caring" and the "cared-for." The "one-caring" is charged with being attentive enough to understand the needs of the other ("engrossment") and with allowing the other's needs to determine what action is appropriate ("motivational displacement"). The "cared-for" is responsible for completing the caring by responding to it in some fashion. This approach is compatible with the Principle of Cooperation, which suggests that cooperation is actuality part of our basic nature. It is possible that both "justice" and "caring" are components of our moral framework.

Gilligan has also proposed that transitions between stages are fueled by changes in our sense of self rather than changes in cognitive capability. In other words, she proposes de-linking moral development from physical maturation. This approach certainly fits with our discussion of the limitations of viewing psychological maturation as a sequence of hierarchical stages. While Gilligan's theory is cast in a series of stages, her framework is compatible with the idea of a development of consciousness (including moral consciousness) that does not necessarily go hand-in-hand with physical development.

If we take the approach of separating physical maturation from moral maturation, we can consider moral development as beginning from our learnings in other lifetimes rather than from a blank slate at our physical birth. We can reflect on how the expression of our actual level of moral development might be limited initially by our level of physical maturation in this lifetime. It is possible that "young souls" are more focused on self-interest; but, at this point in human history, many if not most of the human family will have had an opportunity to mature past self-interest and enter life assuming an attitude of cooperation.

An Alternate View of Maturation and Purpose

We have seen that the traditional models of psychological maturity don't fit well with what we now know of the principles that govern our cosmos and our relationships with others. What if we begin instead from our knowledge that we are all part of a cosmic energy field? We know that we are connected with every other life form through what the cosmologist Brian Swimme calls the "all-nourishing abyss."[12]

In *Principles of Abundance*, we began the exploration of how it is that we can be unique individuals and also be part of a greater cosmic whole, posing the following possible explanation:

"In the beginning," the One Life existed as undifferentiated energetic potential. Choosing to differentiate led to the development of awareness and the possibilities of new learning and growth. Thus, the cosmos is a complex experiment in simultaneous unity and uniqueness, along a continuum of various types of group consciousness. This experiment is going on in a number of settings (galaxies), each of which has different variables at play within the various universes (solar systems).[24]

Within this context, what is our individual developmental task as we learn about how to be both unique and a part of the cosmic whole? Our starting point is that we are energetic entities who have to learn to manipulate a physical body that starts out in an uncoordinated state.

Contrary to the traditional description of the infant fused with its mother, we could posit an infant who is helpless because of finding itself in an entirely new situation within a mechanism (the physical body) that doesn't work very efficiently at first. This infant has come from an environment in which all was known and communicated intuitively to an environment in which its mental abilities are suppressed by physical immaturity.

Experience with "baby signing," or the use of abbreviated American Sign Language with infants, confirms that — contrary to popular stereotypes — infants are actually thoughtful and communicative even when very young.[25] By the age of six months, infants can sign competently, while it takes until they are between 12 to 20 months old before their bodies have matured to the point where verbal communication is clear.

The first major task confronting a new human is indeed individuation, as asserted by a number of psychological theorists, but perhaps for a different reason. Traditional models describe the infant as unable to differentiate self from other, as if this were a failing. Based on the Principle of Interconnectivity, though, we could assume that an infant's apparent self-absorption is simply a task-specific focus as it tries to learn the boundaries of "self" and how to care for "self." We could say instead that, while the infant retains that awareness of interconnectivity with all living beings throughout the cosmos, it now needs to learn how to narrow its awareness to focus on a sense of self that is independent of that interconnection.

In parallel, the infant needs to learn to become socialized into the worldview of its community and to be able to communicate effectively with others. This process involves the acquisition of what Cook-Greuter calls the "language habit,"[26] or the language conventions that shape how we view and deal with the world. Thus, the infant begins to internalize the language habits of its community in order to be able to communicate about shared perceptual experiences, or to see the world as others see it.

The soul may provide a subliminal connection to our experience stream. It can bridge or link between the unique individual and the whole of the One Life since it is part of the universal Soul. If our learning process stretches over multiple lifetimes, the soul would provide us with continuity between those lives. Our soul would thus be the bridge between the incarnate/physical plane and the discarnate/spiritual realm.

One of the confusions that has emerged in the spiritual literature is the relationship between our discarnate and incarnate existences. In order to focus on our physical plane existence, we seem to screen out or forget our actual identity as part of the One Life. Spiritual gurus who have glimpsed or remembered our broader cosmic context speak about our maturational task as attaining a state of bliss in which we can identify primarily with that discarnate existence. This makes sense if we assume that our starting point is physical birth.

But what if it is not? What if our starting point is our discarnate existence from which we choose to incarnate? In that case, we would already know how to be discarnate, part of the cosmic energy. Despite what some spiritual traditions say, we would not need to focus on that or learn that. We already *are* our soul. What we would then have to learn is how to be incarnate as a unique individual and still retain that sense of being part of a whole.

Our self-awareness is necessarily restricted at first. Once we move on from individuation to identification with others and with the One Life, we are in a position to live the Principle of Interdependence. The Ageless Wisdom proposes that "the mode or method of development for humanity is self-expression and self-realization. When this process is consummated, the self expressed is the One Self."[27]

Within the framework of living multiple lives, we can assume that each person enters this incarnation with a particular soul agenda or life purpose—specific lessons to learn and contributions to make. Some may be working on basic coordination—physical well being, emotional detachment, mental focus. Some may be focused on rendering a particular type of service. Some may be consolidating learnings from a particularly challenging former life.

In carrying out our soul agenda, we each have certain strengths on which we can draw. These are givens that are available to us, rather than characteristics that we need to develop, and they differ from person to person. They are based in part on previous incarnational experience and learning and in part on the fundamental skills needed for a given life purpose. It is in this context that we can appreciate the contribution of strength typologies such as the Myers-Briggs and the esoteric model of rays and astrological configurations.

In this context, it becomes important to clarify the "end" or purpose towards which we are preparing as we mature. Some would call this our goal in life, our personal mission. Some would call this our soul's purpose, the contract made before we enter this life. Some would call this our dharma, the work we are called to do because of the karma (positive and negative) we bring with us into this life.

> **Exercise: Identifying One's Life Purpose**
>
> If you already have a good sense of your life purpose, try writing it in 25 words or less, like a mission statement:
>
>
>
> If not, here are some ways to help determine your life purpose:
>
> I feel most "in the flow" when ...
>
>
>
> It is really important to me that ...
>
>
>
> I feel that the following is "mine to do": ...

Our life purpose involves a commitment to becoming the most we can be, to making the contribution that is uniquely ours to make. We become able to see both the concrete/temporal and the eternal/symbolic simultaneously. Our principles tend to shift from being externally defined to being based on our own internal sense of justice, tolerance, respect, and the dignity of all. When most centered with regard to life purpose, we experience a sense of bliss and of being "in the flow."

Developing a Model That Supports Harmlessness

Existing well-known models of psychological development that stop at self-actualization are problematic in that they run counter to the principles under which our cosmos operates. They position us, entering this life, as self-absorbed, competitive individuals concerned only with our own well being. This does not bode well for affirming harmlessness as our core ethic.

Once we consider maturational models that do not assume hierarchical stages and that include a spiritual dimension, we have the possibility of a different perspective that is in keeping with all seven cosmic principles. We can take Teilhard de Chardin's famous quote—"We are not humans having a spiritual experience but spiritual beings having a human experience"—one step further.

ELEVEN

Our Maturational Opportunity

What will it take to mature into harmlessness as a routine and reflexive part of our being? It is clear already that we will have to make some fundamental changes in how we conceptualize maturation because we are not becoming harmless using our current models. In a way, that is not surprising because we have seen the gaps between traditional psychological development theories and the seven principles underlying the operation of our cosmos (which were explained in *Principles of Abundance for the Cosmic Citizen*).

Although many argue for a sequential series of nested stages of development, the Principles of Nonlinearity and Adaptability make it clear that this is not the nature of how we grow. And we can also tell from the level of violence with which we live that our traditional approach is not working. Building in harmlessness is not a matter of adding to, or modifying, the maturational stages that psychologists have theorized to date. What do we need to change so that our comfort with harmlessness blossoms and our attachment to harmfulness withers away? How can we make harmlessness the automatic consequence of our maturation process?

Redefining Our Maturational Goal

Going back to what we know about the cosmos, we know that the very nature of the cosmic energetic field means that there is constant vibratory activity. "All things in the universe are constantly oscillating at different frequencies. These oscillations generate wavefields that radiate from the objects that produce them."[1] Our energetic makeup is such waveform energy packets.

On the biological level, Candice Pert describes receptor molecules as responding "to energy and chemical cues by vibrating. They wiggle, shimmy, and even hum as they bend and change from one shape to another. . . . Basically, receptors function as sensing molecules—scanners. . . . They hover in the membranes of [our] cells, dancing and vibrating, waiting to pick up messages carried by other vibrating little creatures, also made out of amino acids, which come cruising along."[2]

When something vibrates, it is in motion around an equilibrium point. In water, it is easy to see this dynamic as ripples spread out from a central point. The rate of activity, however, varies in terms of the amplitude of vibration. We can think of amplitude as modeling the process that we go through as we are learning a new skill. Through trial and error, we reach the point where we can consistently demonstrate the new behavior.

The biological sciences offer another perspective on growth. Lynn Margulis has shown that bacteria develop through sharing genetic information, or genetic recombination. "As a result of [the ability to routinely and rapidly transfer bits of genetic material], all the world's bacteria essentially have access to a single gene pool and hence to the adaptive mechanisms of the entire bacterial kingdom. . . . The result is a planet made fertile and inhabitable for larger forms of life by a communicating and cooperating worldwide superorganism of bacteria."[3]

If we look at the Ageless Wisdom literature, we find exactly the same descriptions of growth—as vibration, as spiral cyclic motion, and as sharing. "All growth is cyclic and one progresses from step to step in spiral fashion."[4] Growth is pictured as occurring in all directions, not just vertically as is the case in a hierarchical model. We expand in our relationship with ourselves, with others, with our soul and the One Life.

So what can we extrapolate from our cosmic dynamics to help us understand our maturational process? What do they suggest about harmlessness? Think for a moment about a time that you felt caught up in dynamic personal growth. Think of the sense of potential, of possibility. What was that energy like? Probably very resonant with a steady pulsing coming from harmonious vibrations. Now think of a time of anger and upset—what was the energy like? It probably felt jagged and discordant.

If we look at the activity patterns of energy waves, what we find is behavior called interference. As the biologist Bruce Lipton illustrates, when two wave patterns are aligned or in synchronicity, they reinforce each other and produce a more powerful dynamic—known as constructive interference, or harmonic resonance.[5] This is why we feel really good when we are internally centered (in rhythm with ourselves) or we interact with others who are happy and with whom we feel comfortable.

Harmonic resonance also accounts for the so-called "Mozart Effect"—the positive effect on spatial-temporal reasoning of listening to Mozart and other classical or baroque composers. "The effect of classical music on the brain [is] composed of two effects that act in synergy. The first is due to rhythm, which synchronizes with the body's vital rhythms . . . and produces the proper mood for increased cognitive and creative capabilities. The second effect that acts in synergy with the first is melody, which . . . gives to the person the warm feeling that he or she is able to tackle new challenges by setting a path in the invention of new solutions and providing the ability to make the correct choice among possible solutions."[6]

What this means for us energetically is that our waveform energy—our essential nature—has a dramatic impact on ourselves and others. When we are "in synch" with positive dynamics, we intensify them. So we have the potential to focus on harmless, empowering actions and dramatically increase their potency.

Of course, we also have the potential to dramatically intensify the impact of harmful dynamics when we allow ourselves to engage or resonate to those activities. The intensification of energy through constructive interference helps explain phenomena like "herd mentality," "crowd psychology," or "groupthink." It also helps us understand why people are less likely to be helpful in an emergency when there are a lot of bystanders looking on—known as the bystander effect.

When energy waves are out of synch or are not coordinated, the result is known as destructive interference. Here the various wave patterns literally cancel each other out instead of reinforcing each other. We see this dynamic when someone tells a racist or sexist joke and we don't laugh—the joke falls flat. This suggests that one way we can influence harmful behavior is by simply not responding (assuming we can remain safe as we do so). One of the standard instructions to persons being bullied, for example, is to not allow themselves to become provoked and not to respond. This strategy can be problematic both because it places the responsibility for constructive resolution with the recipient, not the perpetrator, and because there are times when being assertive and responding is the psychologically healthy response.

What do these dynamics suggest about our maturation and harmlessness? We know from the Principle of Interconnectivity that we are each indeed energy waves in, and connected through, the vast cosmic energy field. Our whole universe is composed of vibration.[7] Acknowledging our energetic makeup, we can redefine the goal of maturation as *the ability to manage how we focus our energy*. Wave physics shows us that our influence is energetic. We will not have the choice between behaving harmfully or behaving harmlessly unless we do develop that control.

If we accept this definition, we immediately get confirmation of its usefulness because it allows us a vision of maturation that does not involve linear sequential stages. While there may be a need to think in terms of stages in childhood development when there are constraints imposed by the physical maturation process, that is not necessarily the case in adulthood.

So instead of stages, we can focus on the personality dimensions that are essential to gaining this control. We can think of our maturation as multi-directional development. That development, in each instance, is likely to follow a pattern of new insights (perceptual awareness), followed by experimentation in applying those insights (discretionary judgment), followed by a period of integration to consolidate the new learning.

Our very nature is energy waves, and we exist within a single cosmic energy field. In managing our energy, our choice of focus remains an ethical issue. And so a corollary component of our maturation will need to be clarity about harmlessness as our chosen ethic, supported by a set of core values.

Reflecting back on the "Examples of Possible Universal Values" listed in Chapter Eight can help us identify which values are of most importance to us in behaving harmlessly. And those values can guide how we focus our energy.

> **Exercise: Examining Values**
>
> When you need to make a choice, what are the three core values you use in choosing how to act?
>
> 1.
> 2.
> 3.
>
> Give an example of a choice you made that was based on one or more of these values.

As we reflect on the dynamic of harmlessness, we can identify seven key dimensions that are relevant to its practice and that we can strengthen over time: self-discipline, responsibility, decision making, complexity, nurturance, goodwill, and compassion. Each of these seven dimensions is linked to one cosmic principles in particular:

Principle	*Maturational Dimension*
Interconnectivity	Self-discipline
Participation	Responsibility
Nonlinearity	Decision making
Nonduality	Complexity
Interdependence	Nurturance
Adaptability	Goodwill
Cooperation	Compassion

Each of the dimensions offers us the opportunity to consolidate our commitment to an ethic of harmlessness and to embed harmlessness as our habitual way of behaving. So we can describe our maturational process as refining how we focus energy by developing along each of these dimensions.

Maturing in Self-Discipline

Self-discipline is an overarching dimension that is fundamental to our ability to manage how we focus energy, just as our interconnectivity is fundamental to the operation of our cosmos. It is the characteristic that helps us mature away from impulsivity and begin to give thought to the consequences of our actions. It provides the ability to anticipate the impact of our decisions on the cosmic energy network through which we are all connected.

> ### Exercise: Evaluate Yourself on Self-Discipline
>
> 1. If you had planned to finish a task in four hours and suddenly your available time was limited to one hour, how easily could you readjust your approach in order to finish within the hour?
>
> 2. Under what circumstances would you blurt out whatever came to your mind rather than first thinking about the potential consequences of what you are about to say?
>
> 3. What would lead you to act so that you knew you had done the right thing, whether or not family and friends agreed with you?
>
> 4. Pick a time when you were self-disciplined despite difficulties. What made that possible?

As we use willpower to develop self-discipline, we gain an important cognitive or mental oversight function that is responsible for planning and initiating appropriate, harmless actions. We learn to weigh options and consequences before acting—becoming able to identify those decision points where the focus we choose is critical. As part of this process, we learn to value our ability to control impulses.

Self-discipline is also responsible for inhibiting inappropriate, harmful actions. We learn that it is beneficial to delay certain outcomes. We become able to shift our focus and to distract ourselves in order to reduce the perception of delay. We learn to act based on our own inner values and what we perceive to be right, rather than on the threat of punishment, or "shoulds." This includes the ability to go against the group norm if necessary without worrying about group acceptance. And, perhaps most importantly, we learn to screen out messages that we are entitled to act harmfully.

> ### Exercise: Enhancing Self-Discipline
>
> Try this for a week: Each day select one time when you are tempted to think or act in a manner that is not aligned with your core values.
>
> Notice what makes it possible for you to exert the discipline to remain consistent with your core values.
>
> How can you make that decision automatic?

As we mature, we become more able and willing to cope with difficulties and to persist regardless of immediate barriers. We are able to self-regulate and to maintain an inner sense of stability. In other words, we become able "to stay the course" and pick ourselves up after difficulties rather than being devastated by them. One of our challenges is being able to endure through the process of making mistakes as we learn, without beating ourselves up over them.

Each of these self-regulatory activities takes energy. Research has found that our willpower is not limitless.[8] As we exercise self-discipline, we also need ways to rejuvenate in order to regain that willpower. Listening to music, meditation, time in nature—any activity that grounds and centers us will help us rejuvenate.

As we mature, we learn to observe or monitor our behavior, evaluate our performance, and adjust our actions to maximize success. We refine the ability to work within time limits and assess objectively how we are performing. And we strengthen our sense of self-efficacy, or the belief that we can succeed in influencing our energetic focus.

Maturing along this dimension is critical because it helps us screen out the entitlement messages—"I deserve xyz just because I want it"—that stem from a fear of scarcity and that result in harmful behavior. This dimension provides the critical ability to control impulses to act harmfully and instead choose actions that are harmless and that enhance well being.

Maturing in Responsibility

"Nothing strengthens the judgment and quickens the conscience like individual responsibility," said Elizabeth Cady Stanton. Certainly taking control of how we manage energy is a demonstration of responsibility in action. If we turn to the basic cosmic principles, the Principle of Participation is about taking responsibility for the interpretation we place on what we observe — the focus we choose to maintain. As we mature, we learn to make and honor commitments, including only promising what we can and will deliver. As well, we learn to renegotiate commitments so that they remain realistic as circumstances change.

We learn to accept responsibility for our choices and actions, as well as for their effects on others. As part of our learning process, we reflect on the consequences of our choices and modify our behavior as needed in the future. We also learn to accept blame and unpleasant consequences if we do not honor a commitment.

Exercise: Evaluate Yourself on Responsibility

1. How easily are you able to meet a commitment cheerfully when another activity (that you'd rather do) conflicts?

2. What helps you to accept that you cause your own emotions?

3. How willing are you to accept blame for your misconduct even if others are not aware of it?

4. Pick a time when your behavior was particularly responsible, going beyond what was absolutely necessary.

   What made that possible?

We move from blaming others and being defiant to being participants in the co-evolution of the cosmos. This includes a recognition that we cause our own emotional responses by how we interpret our experiences to ourselves — one of essential ramifications of the Principle of Participation.

Part of our responsibility is managing ongoing change. As we come to points in our lives where our past choices and behavior patterns no longer serve us, there is typically some kind of crisis that precipitates an awareness that we need to change. As we mature, fortunately we may choose to change without the angst of a painful crisis.

With or without a crisis, we then need to be able to persist in unlearning old behaviors and replacing them with new behaviors, a process that is often stressful and uncomfortable. If we endure through this conscious learning process, we will eventually reach a point where the new behavior becomes unconscious or automatic and our sense of comfort returns.

Exercise: Enhancing Responsibility

Try this for a week: Each day, notice a commitment that you are making (where you say "I will do . . .") and attach a completion time to it (". . . by x time").

Then notice whether or not you met that commitment. If you did, what helped you do so?

If you didn't, what interfered and how can you change that?

Maturing along this dimension is critical because it supports us in taking our place as a creative part of this interconnected cosmos through the choices we make. Becoming more responsible decreases the likelihood that we will allow ourselves to behave in a harmful manner because we will recognize that we are only harming ourselves by doing so.

Maturing in Decision Making

An important part of accepting responsibility is effective decision making. Before harmless behavior can become reflexive, we need to establish it as a habit through conscious choice.

> ### Exercise: Evaluate Yourself on Decision Making
>
> 1. What helps you to become aware of assumptions or biases so that you can evaluate their relevance to the decision you are making?
>
> 2. If others feel certain that you should make a given decision, what helps you to consider other possibilities?
>
> 3. What techniques do you use to weigh alternatives or scan for options?
>
> 4. Pick a time when you feel you made a difficult and unpopular, yet needed, decision.
>
>   What helped you make the decision?

Successful decision making involves broadening the scope of options that we consider, as well as choosing the appropriate values to help us in decisions. Accepting that time is not linear (the Principle of Nonlinearity) frees us to remake previous decisions that didn't serve us, thus healing the "past."

As we mature, we shift from being reactive to being able to proactively determine the direction of our development. This requires that we become aware of both explicit and implicit assumptions or biases so that we can examine their validity and not be inappropriately influenced by them. This also helps us increase our willingness to take risks and fail, and to ask forgiveness rather than permission.

Exercise: Enhancing Decision Making

Pick a decision you need to make. Select a possible solution and live for several hours "as if" you had selected that solution.

Then select another solution and live "as if" you had picked that option, and so on until you have tested each alternative.

What did you learn about your decision making process?

The concept of "metacognition" refers to how we think about ourselves and our behavior, including how we make ethical judgments.[9] "For example, I am engaging in metacognition if I notice that I am having more trouble learning A than

B; if it strikes me that I should double check C before accepting it as fact."[10] This skill set is directly related to the decision making that controls our energetic focus and results in harm or harmlessness as we evaluate and revise our own thinking processes and subsequent choices.

As we identify and evaluate options, we are brought back to our core values. Are we concerned primarily with selecting the option that is beneficial to the most people? That harms the fewest people? That treats everyone involved fairly? That upholds a particular universal value?

We also learn when to consult with others before deciding and when, with issues that are time sensitive, we need to just decide ourselves. While often our decision making process is logical and objective, we learn to trust our intuition, especially in situations with more time pressure or increased ambiguity.

Once we have chosen an option, we need not only to implement it but also to evaluate its consequences as part of our ongoing learning regarding decision making. This commitment to evaluation can help us in making a difficult choice as we are usually able to choose an option for a limited time period, subject to re-examination. It also involves a willingness to limit, or commit, ourselves to one particular course of action rather than trying to keep all of our options open.

As we gain practice in making conscious decisions, we develop the ability to anticipate consequences in the future and base decisions on factors other than anticipated immediate outcome. In some societies, we are expected to become able to anticipate consequences for seven generations!

Maturing along this dimension is critical because it supports our being able to shift from choosing the immediate pleasure to more sustainable options. We become able to make harmlessness, and the energetic focus that supports it, our habitual choice.

Maturing in Complexity

We have already seen that one of the central characteristics of growth and maturation is increasing complexity. As organisms that absorb information and interact with our environment, we are engaged in continuous learning opportunities. So an important part of managing our focus is being able to manage complexity.

> **Exercise: Evaluate Yourself on Complexity**
>
> 1. How careful are you to identify more than two possible options?
>
> 2. When you are feeling uncertain, how easy is it for you to seek out additional information rather than moving quickly to a decision?
>
> 3. How easy is it for you to help someone else generate a range of options even if you have one particular option to which you feel attached?
>
> 4. Pick a time when you were particularly successful at "both/and" or "win/win" problem solving.
>
>   What made that possible?

As we mature, we move away from a simplistic good/bad, right/wrong, either/or perspective on the world. This dimension is critical to our willingness to disengage the "us-them" dichotomy that generates violence and is antithetical to the Principle of Nonduality.

We shift from selecting only those aspects of a situation that agree with our own perceptions to considering a broader range of options. It becomes more difficult for us to be certain that one position is all "right" and another position is all "wrong." As we learn to perceive nuances and subtle differences that might otherwise be overlooked, we realize that there is usually a continuum of alternatives, not simply two extremes. We become more able to seek multiple alternatives and engage in "both/and" problem solving.

Exercise: Enhancing Complexity

Try this for a week: Each day pick a time when you find yourself making a quick decision just to get it made and ask yourself to name two other decisions you could make instead.

Then reflect on how the outcome might have been different if you had chosen one of the other options.

What did you learn about managing your tolerance for ambiguity?

An important aspect of this dimension is becoming better able to tolerate ambiguity and paradoxes, remaining open to input for a time rather than pushing for immediate closure and certainty. Tolerance for ambiguity is critical if we are to consider new options that have not yet become clear or are only working hypotheses. If we cannot tolerate a reasonable amount of ambiguity, we are likely to find the lack of clarity or structure uncomfortable and even threatening.

Instead we need to be able to recognize that answers are not always straightforward and that there may be more than one possible solution. We learn to consider and test out multiple meanings or perspectives. Maturing along this dimension is critical in being able to multi-task without losing our focus and being able to manage ambiguity in a constructive manner. It allows us to consider a range of options rather than viewing the world simplistically and mechanistically.

Maturing in Nurturance

Nurturance is the dimension that supports our ability to empower ourselves and others. Here we broaden our focus from ourselves to our cosmic context. We can think of this like a figure-ground reversal—we as individuals go from being the "figure" of primary focus in our lives to being part of the group "ground." Our sense of being part of a whole and related to, rather than separate from, is essential to our understanding not only of the Principle of Interconnectivity but also the Principle of Interdependence.

This shift in focus brings with it a concern about the needs and priorities of others. We learn to tolerate and then respect multiple points of view, becoming less likely to judge others. We come to recognize and repudiate ways in which we use others to meet our own needs instead of attending to theirs.

Being nurturing is a responsive and receptive way of relating to others, complementing the self-determined approaches of self-discipline and decision making. Respect for others is a critical attribute, leading to careful listening and a desire to empower others to speak for themselves rather than speaking for them.

> **Exercise: Evaluate Yourself on Nurturance**
>
> 1. How eager are you to hear and understand another's point of view if it differs from your own?
>
> 2. What helps you not get overly critical if you make a mistake?
>
> 3. What helps you be supportive of another person while staying centered within yourself?
>
> 4. Pick a time when you were able to be respectful of a position very different from your own.
>
>   What made that possible?

Coupled with respect is a genuine curiosity about others and what their lives are like, wanting to understand multiple points of view. Nurturance leads us to value diversity and recognize and enable the contributions of all beings. We seek out win-win strategies and welcome alternate outcomes without being attached to our own preferences. We feel compelled to ensure universal human rights and advocate for those of all life forms not able to advocate for themselves.

> **Exercise: Enhancing Nurturance**
>
> Try this for a week: Each day select someone who you would normally get irritated by.
>
> Pay attention to them and try to determine what is important to them.
>
> Put yourself in their shoes until your empathy washes away your irritation.
>
> How can you make that empathy more likely than irritation as your automatic response?

Maturing along this dimension is critical in ensuring that we take the needs and priorities of others into consideration as we make choices about our energetic focus. Nurturance is central to the exercise of harmlessness because it keeps us from degrading and dehumanizing others, which is the first step towards violence.

Maturing in Goodwill

While nurturance assumes an active engagement with a particular person, goodwill is a more detached attitude towards both individuals and groups. Goodwill refers to our willingness to value others, even if their perspective is quite different from our own. It reflects our general willingness to release self-interest and identify with the good of the whole. It serves to empower ourselves and others since it encourages us to emphasize points of commonality.

> **Exercise: Evaluate Yourself on Goodwill**
>
> 1. What helps you to affirm what you want while releasing how it might occur?
>
> 2. What helps you to release the need to feel "right" or that you have the "right" answer?
>
> 3. What helps you to set an issue to the side while you wait for events to unfold rather than pushing for immediate closure?
>
> 4. Pick a time when you consciously supported another person in being successful rather than doing it yourself. How did you feel? What helped you release any attachment to your own recognition?

Goodwill is the dynamic that generates a longing for harmony and cooperation, a wish to free ourselves from any attachment to hatred or revenge. It is grounded in detachment, or the ability to interact with others without being vested in how they behave or the particular choices they make. It allows us to respect multiple points of view, though we may choose to act as a catalyst by challenging another's perspective.

A sense of fairness is a part of detachment and leads us to ensure the same outcome for everyone, without playing favorites or giving special deals to a select few. While special circumstances may call for flexibility, that flexibility would apply to anyone regardless of their personal status or connections.

> **Exercise: Enhancing Goodwill**
>
> Try this for a week: Pick an hour each day when you consciously view everyone you see or interact with in a positive light rather than judging how they "should" be.
>
> What did you learn about being both respectful and detached?

Historically, goodwill has been viewed primarily as a spiritual practice. But it is linked to a range of altruistic behaviors such as charitable giving. It reflects an expansion in our range of concern to include others in it. Interestingly, research has found gender differences in this dimension. Of those exhibiting goodwill, women were more generous with their resources and were equally generous to those similar to (in-group) and different from (out-group) themselves. Men, on the other hand, were generous primarily to in-group members.[11]

Releasing self-interest and being willing to focus on the best possible outcome for all can be challenging. We tend to personalize other people's responses. There are a number of techniques that can help us to release emotional attachments and hold an open, non-judging focus. One such technique is "focusing," developed by psychologist Eugene Gendlin, which helps us become aware of what we feel or want in a detached manner.[12]

Maturing along this dimension is critical because we are able to see that others have their own life agenda, rather than seeing them as a threat to ourselves. Goodwill allows us to focus supportive energy without forcing a particular outcome, maintaining a nourishing context where the harm reflex is not triggered.

Maturing in Compassion

As we mature, we begin to move beyond a preoccupation with ourselves and to understand that others have their own issues and perspectives. We shift from self-focus and a "what's in it for me" attitude to a curiosity about others' experiences and then to a cosmic perspective. We begin to comprehend what another is feeling and become increasingly willing to assume another's good intentions.

> **Exercise: Evaluate Yourself on Compassion**
>
> 1. How are you best able to be compassionate with yourself?
>
> 2. What helps you shift from expecting perfection to recognizing that we are all learning?
>
> 3. What helps you recognize another as a soul?
>
> 4. What helps you distinguish between a person (as a soul who doesn't intend harm) and the person's actions (which may be harmful)?
>
> 5. Pick a time when you were really upset with yourself. Viewing yourself in retrospect with compassion, why did you become so upset?

Part of this maturational dimension is our growing willingness to forgive ourselves and others, to focus on what we are learning rather than on how we have failed. We come to realize that it is impossible to grow without experimentation and making mistakes, and so we begin to cultivate gentleness with ourselves and others. When errors occur, we become able to feel regret and sadness rather than anger or blame.

We develop true compassion by beginning with ourselves, accepting and loving ourselves without judgment or evaluation. We extend this attitude to others, including other life forms, recognizing that all beings deserve kindness and understanding. In short, we become more thoughtful of others and able to walk in another's shoes.

> **Exercise: Enhancing Compassion**
>
> Try this experiment: Pick someone you don't get along well with. Imagine yourself into their life.
>
> Why might they act the way they do?
>
> What are they likely to be thinking or feeling?
>
> How does this shift the way you experience them?

Maturing along this dimension is critical because it triggers a longing for the good of the whole. Compassion is a key component of harmlessness. The stronger our ability to be compassionate with ourselves and others, the more likely it is that we cannot bear to inflict harm.

Ensuring the Habit of Harmlessness

What we have just traced is a maturational shift in how we focus our energy. This shift moves us from simplistic, impulsive behavior where we play a largely passive role to the proactive ability to focus on what will be most empowering for the well being of us all.

The psychological development model being proposed is that we mature on each dimension at our own pace, starting from the strengths that we brought with us into this life. Our maturation process is neither linear nor sequential. We may focus on all dimensions simultaneously or on only a few. We make progress in multiple directions, consolidating and extrapolating as we go.

It is these maturational dimensions that help us develop the "reflex" or habitual focus necessary to implement an ethic of harmlessness. And they provide the discipline necessary to live joyously and successfully as part of the cosmos and to achieve our life purpose.

TWELVE

Developing an Ethic of Harmlessness

We live in a time of great personal and social potential. We are at the tipping point where we can now choose to move beyond a constant undercurrent of harm and instead embed the habit of harmlessness as our expected and usual response.

In 1993, the Parliament of the World's Religions, with participation from over 100 spiritual traditions, adopted the "Declaration Toward a Global Ethic."[1] It asserts, in its Introduction: "We affirm that there is an irrevocable, unconditional norm for all areas of life, for families and communities, for races, nations, and religions. . . .We are interdependent. . . . We take individual responsibility for all we do. . . . We make a commitment to respect life and dignity, individuality and diversity so that every person is treated humanely, without exception."[2]

Public position statements like the Declaration Toward a Global Ethic are important supports for our moving forward together towards an ethic of harmlessness. We have seen how the Butterfly Shift can provide us with glimpses of what it would be like to live harmlessly on a permanent basis. We have developed a new definition of our goal for psychological maturity—*the ability to manage how we focus our energy*—that can help us with "how" we would implement a global ethic. Now we need to explore the characteristics of a global ethic that could ground us in harmlessness and then determine how we could monitor our progress towards achieving that goal.

Reclaiming Harmlessness as a Strength

We noted, in Chapter Two, that harmlessness is sometimes viewed as a weakness—as being unable to have any impact. In actuality, harmlessness requires significant self-discipline, as we shall see. It also requires a commitment to five types of actions:

1. *Refraining from acting harmfully*

   At its most basic, harmlessness involves a willingness to keep from behaving harmfully towards self or others in thought, word, and action. The most challenging situations for us are likely to be in the arena of thought—disciplining ourselves to eliminate negative or judgmental thoughts about others and ensuring that we do not engage in self-deprecating or demeaning self-talk. We may also have challenges in making sure that we take good physical care of ourselves.

   The most challenging interaction patterns to change are usually our everyday "mindless" routine interactions with persons that we tend to take for granted. These are the situations with which the Butterfly Shift can help us practice a change.

2. *Bearing witness to the harm that has been done to others*

   There are some situations in which we hear of harm that has been done even though it is too late or not possible for us personally to intervene. In such situations, we can at least bear witness—recognizing that it has occurred, honoring the fortitude of the recipients, becoming sensitized to the circumstances that allowed it to occur, and alerting others. For example, Rubina Feroze Bhatti of Pakistan has made a moving documentary film about the illegal practice of wan'ni (sometimes called vanni) that she shows repeatedly in villages where the practice is still pervasive.³ *Wan'ni: Murdered Marriages* depicts the lives of young rural Pakistani women who have been ordered into forced marriage by village elders as restitution to families that have suffered at the hands of their male relatives. In some cases, the crime committed by the male relative was murder, and the male perpetrator goes free while the young woman serves a life sentence as a virtual slave in a household where she is despised and abused.

Some of us may want to avoid the responsibility of knowing that such violence is occurring. However, the legal concept of "willful blindness" holds persons accountable even though they have successfully avoided gaining detailed information about crimes that they knew were to be committed. We can also harken back to Nuremberg Principle VII, which holds persons legally responsible when they have been complicit in the commission of a crime. With today's Internet and ready access to information, it is difficult to say truthfully that we do not know that violence is occurring worldwide.

3. *Empowering those who have been harmed*

One important action that we can usually take is being caring and compassionate with the recipients of harm and assuring them that it was not their fault. Our challenge may lie in both holding the recipient blameless and not expecting the recipient to resolve the situation. The perpetrator needs to be held accountable or, failing that, we need to take responsibility to make sure the harm is not repeated.

4. *Intervening reactively*

Reactive intervention involves taking steps to stop harm while, or just before, it occurs rather than turning aside and ignoring what is happening. Direct opportunities for intervention are likely to be the most obvious, though what we should do may not be so obvious. Intervening to prevent indirect harm is less obvious. Such situations can include speaking out when others (who are not present) are being demeaned so that negative energy about them is not generated. For example, when we hear someone say something untrue ("Marnie isn't really qualified for the job") or unflattering ("Dave's gotten so full of himself—he just ignored me the other day"), we could intervene to set the record straight ("Actually Marnie has ten years of experience as . . .") or provide an alternate, nonjudgmental explanation ("I hear Dave's father is very ill . . . perhaps he's been preoccupied.").

Exercise: Proactive Harmlessness

Pick one recent instance in which you did *not* intervene proactively when another was being harmed.

Why did you refrain from intervening?

How did you feel about your choice?

Now pick a recent time in which you *did* intervene proactively to prevent harm being done.

Why did you decide to intervene?

How did you feel about your choice?

5. *Intervening proactively*

Proactive intervention is really what positive harmlessness is all about. It includes a willingness to prevent others from being harmed by individuals or groups, including society at large. It means not only resolving matters in an immediate situation but also identifying and addressing the root cause so that future harm is avoided.

The global social action network, Avaaz, has set an impressive record for proactive initiatives to combat violence worldwide.[4] Since its beginning in 2007, it has grown to over 4.9 million members from every nation, operating in 14 languages. It provides a platform for ordinary citizens to join together and successfully lobby political leaders on issues of concern, thus ensuring "that the views and values of the world's people shape global decision-making." Most recently, Avaaz has been successful in blocking plans by Tanzania and Zambia to resume killing elephants for their ivory tusks, in helping ensure that the U.K. will act to protect oceans, and in raising funds to fight the rape trade in women and girls.

Some say that we have no right to judge others' actions when cultural values different from ours appear to be involved. Most commonly, this position is taken when women are being abused under traditional customs. But are we truly to ignore common occurrences like women being stoned to death for tripping and inadvertently showing their ankles and lower legs, prepubescent girls being sold as brides to older men to pay off their father's debts, women being horribly scarred with acid or burned to death because their husbands were displeased with them? The 2003 Nobel Peace Prize Laureate Dr. Shirin Abadi was quite eloquent on this point, as a Moslem woman, when accepting the Nobel Peace Prize:

> The discriminatory plight of women in Islamic states ... whether in the sphere of civil law or in the realm of social, political, and cultural justice, has its roots in the patriarchal and male-dominated culture prevailing in these societies, not in Islam. This [patriarchal] culture does not tolerate freedom and democracy, just as it does not believe in the equal rights of men and women, and the liberation of women from male domination (fathers, husbands, brothers) because it would threaten the historical and traditional position of the rulers and guardians of that culture.[5]

The 1948 UN Universal Declaration of Human Rights clearly states that all persons—regardless of the culture in which they live—are entitled to dignity and safety of person.[6] To make this principle clear in relation to women and violence, the UN passed the 1993 Declaration on the Elimination of Violence Against Women.[7] These two Declarations constitute part of the framework of international law for the 192 Member States of the United Nations, which include Islamic States such as Afghanistan and Saudi Arabia that have had the most widely-publicized gender discrimination and gender violence. The issue is not sensitivity to cultural differences but rather whether or not UN Member States are to be held accountable for gender violence by the international community.

An ethic of harmlessness requires the tension of remaining open and respectful while also being alert for opportunities to intervene. Clearly this is not an easy or effortless position.

From Reciprocity to Harmlessness

The Declaration Toward a Global Ethic is only one of several initiatives launched to counter the rising tide of violence.

Hans Küng, author of the Global Ethic, describes it as "the necessary minimum of common values, standards and basic attitudes," stating that it requires "a minimum basic consensus relating to binding values, irrevocable standards and moral attitudes, which can be affirmed by all religions despite their undeniable dogmatic or theological differences and should also be supported by non-believers."[8]

Exercise: Harmlessness and the Golden Rule

Keeping in mind the concept of harmlessness as a strength, list three ways in which you believe following the Golden Rule would result in your behaving harmlessly:

1.
2.
3.

Now list three ways in which choosing to be proactively harmless might lead you to behave differently than if you chose the Golden Rule as your ethical standard:

1.
2.
3.

What do your answers suggest to you about how you view harmlessness?

The movement toward the adoption of a common global set of ethical standards has coalesced around the ethic of reciprocity, or the Golden Rule—often stated positively as "Do to others what you would wish them to do to you." In addition to the proposed Global Ethic, we have already reviewed (in Chapter Six) the Charter of Compassion, which also references the Golden Rule. Similar initiatives, under the auspices of the Interfaith Peace-Building Initiative of the United Religions Initiative, include a Golden Rule Poster, a Golden Rule Resolution, and the declaration of April 5 as Golden Rule Day.[9]

In September 1997, the InterAction Council proposed "A Universal Declaration of Human Responsibilities," drafted as a complement to the Universal Declaration of Human Rights to balance freedom with responsibility. That Declaration is also focused on the Golden Rule, but cited in its negative version: "What you do not wish to be done to yourself, do not do to others."[10]

Claiming an ethic of reciprocity (the Golden Rule) as the ultimate ethical standard draws on familiar language and concepts. This helps us to move forward on a trajectory away from violence. However, the ethic of reciprocity itself has certain limitations if we are to use it to create a milieu of harmlessness in our everyday lives.

First, the functioning of an ethic of reciprocity depends on relationships themselves being reciprocal and of equal power. In actuality, there are a number of types of relationships that are non-reciprocal. The most obvious is that of child and adult. Especially with young children, it would not be appropriate to treat a child in the manner that an adult wishes to be treated, other than in the very broad sense of treating the child with respect. Children require and deserve a degree of care and protection that is not appropriate with most adults.

Tenets of the
Universal Declaration of Human Responsibilities[11]

If we have the right to:	Then we have the obligation to:
Life	Respect life.
Liberty	Respect other people's liberty.
Security	Create the conditions for every human being to enjoy human security.
Partake in our country's political process and elect our leaders	Participate and ensure that the best leaders are chosen.
Work under just and favorable conditions to provide a decent standard of living for ourselves and our families	Perform to the best of our capacities.
Freedom of thought, conscience, and religion	Respect others' thoughts or religious principles.
Be educated	Learn as much as our capabilities allow us and, where possible, share our knowledge and experience with others.
Benefit from the Earth's bounty	Respect, care for, and restore the Earth and its resources.

Persons who have been incarcerated for an offense, and consequentially have had some of their rights restricted or removed, likewise are in a non-reciprocal relationship with authorities within the penal system. And there are many professional-client relationships with a built-in power imbalance in which professional codes of conduct mandate specific non-reciprocal behavior by the professional in order not to overstep power boundaries.

This brings us to the *second* issue, which is the definition of the "other" to be treated reciprocally. In most discussions of the Golden Rule, the examples given are among adult humans. But what about other life forms with whom we exist interdependently (the Principle of Interdependence)? Are we to treat insects or wild animals or flowers or trees as we would wish them to treat us? What would that mean exactly? And what about circumstances where there appears to be no reciprocal "other" because that "other" is viewed as property — as may be the case with women and children in some jurisdictions — to be disposed of as the owner wishes? How do we ensure that that "owner" becomes aware that reciprocity applies?

Third, an ethic of reciprocity assumes that we have a clear and mature perspective on what is in our own best interest. What then of the stance within a gang of "I'm tough . . . I can take it . . . bring it on." Does a person's willingness to be treated violently justify either our doing so or their treating us in like fashion? Or what about the way in which we care for our health — does our desire to overeat on sweets, for example, mean that is it good for others to help us indulge those cravings? There is a growing concern, particularly in the U.S., that obesity is rampant and brings with it potentially serious health problems.[12]

Fourth, an ethic of reciprocity supports the status quo without questioning harmful dynamics within it. Adhering to an ethic of reciprocity allows structural violence[13] to remain in place as well as the presuppositions and language habits that support harmful behavior.[14]

For example, the men of the Dugum Dani (in New Guinea) practice ongoing highly ritualized warfare among neighboring villages with strict rules of conduct that result in little bloodshed and minimal chance of death.[15] It is a way of life with which the men in all the villages are very comfortable—they are happy to treat each other in this fashion. It allows them to display courage and also excuses them from most of the mundane tasks of daily living since they are busy making weapons, preparing to fight, fighting, and dealing with any resulting injuries. But what of the women who bear the brunt of the work to support family and community life? And the young girls who have their fingers cut off with a stone axe to pacify family ghosts as part of funeral rites? What do ritualized warfare and ceremonial violence teach young people about peaceful methods of conflict resolution? Anthropologist Karl Heider, who assisted with preparing the movie *Dead Birds* about the Dani, has called the Dani "peaceful warriors"—an interesting contradiction in terms.[16]

Finally, an ethic of reciprocity is ultimately based on fear. We behave ethically in order to trigger what we want from others, which we are afraid we might not get otherwise. Or we act in a certain way in order not to incur the wrath of another. If we are to become positively harmless, we must be motivated instead by goodwill and a sense of abundance to be shared joyously.

Harmlessness is not simply a more extreme version of reciprocity. It is a qualitatively different dynamic. The impulses of compassion and gratitude that might underlie behaving in accordance with an ethic of reciprocity certainly are a component of harmlessness—as we have seen in the Butterfly Shift.

But harmlessness requires practice and maturity in how we focus and use the energy that flows among us as part of our interconnected cosmic energy field. And if we accept, at least as a working hypothesis, the concept of continuity of consciousness, then surely we can expect more of ourselves than simple reciprocity. Surely we can build a "harmlessness muscle" over our lifetimes that allows us to be positively and proactively harmless.

Bridging the Gap Between Theory and Practice

Simply choosing to live harmlessly is not enough. Our lives are so permeated with habits of harm that we might not even be aware of which habits we need to change. How, then, do we make the shift from harm to harmlessness?

In order to make the changes necessary to embed an ethic of harmlessness, we need to rely on the same three steps we described in Chapter Three for the Butterfly Shift mini-immersion: awareness, emotional engagement, and action. The difference is that, instead of a brief, mini-immersion, we need to focus on a longer term dynamic.

1. *Awareness*

   We cannot change until we become conscious of the myriad ways in which harmfulness is embedded in our lives. These include not only how we behave towards others but how we treat ourselves—including positive self-talk, good physical care, and limiting exposure to unedifying material.

   The language that we choose to use plays an active role in shaping our worldview and the options we consider. We need to become conscious of the words we choose—not only to avoid pejorative meanings, but also to avoid the violence they condone. For example, do we really have to describe visitors to a website as "hits"?

   While we have referenced the UN Universal Declaration of Human Rights multiple times, how are young people to become aware of these rights? Is this Declaration taught in our schools? Is it discussed in social or religious gatherings? What attempts do we make within our communities to bring this universal standard alive by measuring our actions against it?

And we need to shift what we teach ourselves and our young people about who we are. As we explored in *Principles of Abundance for the Cosmic Citizen*, we are interconnected, interdependent, cooperative energetic beings. As such, we are directly responsibility for learning as early as possible how to manage our energy in a harmless manner. We face the challenge of creating a shared worldview based on harmlessness—a context that has not existed before—in order to shift our shared expectation of what is possible.

2. *Emotional engagement*

Commitment to a new way of being with ourselves and each other requires the energy of positive emotion, the "muscle" that we talked about in Chapter Six. We have plenty of experience that tells us that harmlessness can exist in tandem with harm unless we are very careful to eliminate that harm. After all, harmfulness is our current life context!

Emotional engagement requires not just interest in but passion about shifting to an ethic of harmlessness. We need to feel an urgency about the changes needed . . . *all* the changes needed. Immersion experiences carry us part of the way, helping us create a foundation of familiarity from which we can bridge. But we also need to actively build on that foundation until harmlessness becomes our automatic reflex.

3. *Action implementation*

The philosopher Edmund Burke is quoted as saying: "All it takes for evil to flourish is for good men to do nothing." We have plenty of evidence of the evil or harm flourishing around us. Our challenge is what will we do about it? Empowering ourselves means increasing our capacity to make respectful and constructive choices and then transforming those choices into actions and desired outcomes.

Learning to embrace being harmless ourselves is certainly a tall order. But we are called, if we choose an ethic of harmlessness, to do more than make changes in our personal lives. We are called to heal the world from the violence that has been, and is still being, perpetrated.

If we choose an ethic of harmlessness, we have responsibility for ensuring that harm becomes impossible to tolerate. We need to create a social context in which purchasing a child online for sex play, for example, is unthinkable and where the current active traffic in young sex slaves withers away for lack of purchasers.[17]

There is already a growing number of initiatives focused, consciously or unconsciously, on building a context of harmlessness. Peace initiatives, non-violent communication and win-win negotiation skills, gratitude and forgiveness practices, random acts of kindness, and pay-it-forward concepts all contribute to a shift in consciousness in the direction of harmlessness. The portals *Supporting Spiritual Development* and *Values-Conscious Business* provide links to hundreds of such initiatives.[18]

In shifting from a theory of harmlessness to its practice, we also need to be aware of why persons might resist such a shift so that we can find ways to address those concerns. Here are a few of those reasons:

o People may be uncomfortable with any change, with moving away from the familiar even though those familiar habits are problematic. Often this is because they fear the consequences of making mistakes, which will inevitably happen if they are trying new behaviors. The key here is shifting from an emphasis on perfect performance to valuing experimentation and welcoming feedback.

- People may be convinced of the myth of scarcity and be fearful that, if they are not aggressively focused on their own self-interest, they will lose out. The key here is to be able to shift from self-absorption to an appreciation of, and engagement with, others—a kind of figure-ground reversal.

- People may feel self-protective, hurting others before they can be hurt themselves. They may have learned to express their anger or insecurity through violence. The key here is to find ways to build self-esteem around the strengths of positive harmlessness.

- People may define themselves as important to the extent that they dominate and demean others. And they may come to believe that violence is an acceptable way to control an intimate partner or family member. The key here is to shift from an externally defined sense of self to an internal sense of doing what is right.

Exercise: Challenges of Harmlessness

Pick a time recently when you chose to behave harmfully.

Why did you make that choice?

What would help ensure that you were able to choose a harmless response instead if faced with the same circumstances?

o People may define themselves not only in terms of the ability to dominate others but also by being part of the toughest and most violent group . . . the "real man" image that underlies gang violence. The key here is providing different role models that demonstrate that the "real man" is compassionate and caring. This change is intimately linked to shifting from viewing harmlessness as a weakness to recognizing harmlessness as the greatest strength.

While it is useful to gain an understanding of the dynamics of others, our primary responsibility lies with ourselves. What choices are we modeling for others? Are we prepared to say "no more" with regard to harmful behavior?

Measuring Our Engagement with Harmlessness

If we are to undertake the changes needed to embed an ethic of harmlessness, we need to have a clear picture of the different ways in which harm and harmlessness are expressed. The Harmlessness Scale™ below describes eight attitudes, ranging from the extreme of Brutality to the proactive attitude of Advocacy, that capture the complexity of our responses to harm and harmlessness.

Being aware of different attitudes towards harmlessness is not enough, though. We also need to understand where we personally are starting from so that we can identify the changes we might wish to make. By measuring how engrained harm is in our daily lives, we can heighten our sensitivity to habitual harmful responses.

Appendix C contains a series of questions to measure how likely we are to exhibit each of the eight attitudes. If you are curious about how you would rate on harmlessness, you might want to answer those Harmlessness Scale™ Questions before you read about the eight attitudes described below.

The Harmlessness Scale™ Questions are intended for use by individuals, by trainers, and by researchers.

Harmlessness Scale™

Attitude	Characteristics
1—Brutality	Enacting physical violence that allows for no obvious safe haven
2—Harassment	Engaging in verbal and psychological violence, including treating others as commodities
3—Dismissiveness	Treating others, or oneself, as being of no account and not to be taken seriously
4—Defensiveness	Refusing to admit that violence is occurring or assuming that it is justified
5—Abstention	Refraining from being harmful but refusing to intervene to prevent harm from occurring
6—Supportiveness	Recognizing that harm is being done and intervening, but privately to express empathy & concern
7—Nonviolence	Exhibiting zero tolerance for jokes & casual harm; intervening to prevent specific harm
8—Advocacy	Transforming situations that traditionally call for violence into win-win solutions, and dismantling structural violence

The first four attitudes of the Harmlessness Scale™ encompass the ways that we choose to behave harmfully:

1. *Brutality*

    This attitude is the most extreme form of harm, involving overt physical violence to self or others. It includes murder, torture, rape, and incest as well as the "everyday" actions of yelling, hitting, and generally allowing oneself to release tension by battering someone else. It also includes extreme abuse towards oneself, usually intended to help oneself feel better, such as cutting, substance abuse, or high-stress expectations of working long hours with little sleep.

    The common thread here is a lack of an obvious safe haven and the real possibility of one's life being forfeit. Brutality is typified by the philosophy of "an eye for an eye." When tolerated, brutality may be excused ("a husband is king in his home" or "soldiers have a right to relieve their tensions") or even glorified (as in the case of "honor killings").

2. *Harassment*

    This attitude involves the more subtle verbal and psychological forms of terror. It includes belittling and demeaning oneself or another person and generally creating an energetic field of negative thought forms. It also includes the buying and selling of persons and generally treating them like commodities.

    Research has shown that these types of psychological violence are now even more prevalent than physical violence.[19] Harassment can be as permanently damaging as Brutality. One difference is that there is at least a possibility with Harassment of having a safe haven within a group of like-minded persons who are all experiencing the same dynamic.

3. Dismissiveness

This attitude involves treating the recipient as being of no account or not worthy of notice—including the practices of treating persons only as stereotypic members of a group and of shunning.[20] Dismissiveness is often reflected in the choice of language that is used—for example, referring to a grown man as "boy" (as may happen in a racist context) or referring to a grown woman as "girl."

> **Exercise: Dealing with Dismissiveness**
>
> Of the various forms of harmfulness, Dismissiveness is particularly challenging. What harmless responses could you give to the following comments:
>
> 1. "I think Lydia would do better on her doctoral research if mentored by a girl."
>
> 2. (ignoring the elderly woman who is asking about an iPhone and addressing the younger woman with her) "Your mother might find this phone her best choice because . . . ."
>
> 3. (ignoring the woman clerk nearby whose name tag says Electronics Department) "Let's ask that clerk over there. He probably knows how this DVD works."
>
> 4. (speaking to the wheelchair attendant while ignoring the passenger in the wheelchair) "We'll be boarding her in about 20 minutes."
>
> 5. (referring to two Ph.D. scientists) "I'd like to introduce our speakers for this morning—Dr. Ernest Gallawy and Miss Debra Noring."

When the harm originates with a third party, Dismissiveness involves ignoring the violence that is occurring since attending to it would be inconvenient. It also includes a lack of belief in the capabilities of the recipient. Being treated as essentially invisible or of no real worth is very psychologically damaging and demoralizing.

4. *Defensiveness*

In this attitude, a person recognizes on some level that harm is being done but refuses to admit its consequences and the responsibility of the perpetrator. The individual may feel terrible that the harm is taking place but is too fearful to intervene to prevent it. Typically, the consequences of the harm are excused either by justifying it ("well, boys will be boys") or trivializing it ("what's the big deal?") or blaming the recipient ("she was really asking for it").

One such example was the justification given by Angeles Chanon, the mother of one of the teenagers charged with bullying Phoebe Prince who subsequently committed suicide: "'My daughter didn't do any of those things . . . My daughter's not a violent kid.' While she [Chanon] acknowledges her daughter was once suspended from school for verbally abusing Prince, a new student from Ireland, Chanon insists her daughter never harmed Prince physically or urged her to hurt herself."[21] Clearly a misunderstanding about the meaning of "harm" is in operation in this justification as Chanon appears to restrict the notion of violence to physical abuse and to overlook verbal abuse even though the school suspended her daughter for that behavior.

The other four attitudes range from an absence of harmfulness to the practice of "positive" harmlessness:

5. *Abstention*

   In this attitude, the person refrains from initiating harmful behavior themselves but does nothing to intervene when another is being harmed. The general attitude is "it's not my problem" or "it's not my battle." The inactivity results from a fear of the consequences rather than the sneaking feeling that the harm was justified (as is the case with Defensiveness). In essence, this is an attitude of non-engagement.

6. *Supportiveness*

   In this attitude, the person recognizes that harm is being done and will intervene, but only to privately remonstrate with the perpetrator and console the recipient. From a positive perspective, the person does provide general support and encouragement to the recipient.

7. *Nonviolence*

   In this attitude, the person both actively refrains from harmful behavior and moves to intervene in specific situations to stop another from being harmed. This includes bearing witness to the fact that harm has occurred and exhibiting zero tolerance towards harm of any type. When another must be stopped or restrained, it is done with compassion and with the desire to achieve a nonviolent solution.

   Our biggest challenge is how to intervene in instances of Brutality without putting ourselves or the one being brutalized unnecessarily in harm's way. We need practice in creative alternatives. And we need a commitment from communities, particularly schools, to teach young people the options and benefits of responding nonviolently and to provide them with the skills for identifying areas of common interest so that they can negotiate win-win solutions.

> **Exercise: Nonviolence Without Harm to Self**
>
> Since our goal is no harm to others or to ourselves, it can help us take appropriate action if we think of possibilities ahead of time. Suppose you happen upon someone being badly beaten. Some examples of what you could do to be helpful without being beaten up yourself are:
>
> o   Use your cell phone to call 911 for help.
> o   Film what is happening (for evidence) after calling for help.
>
> What other options can you think of:
>
> 1.
>
> 2.
>
> 3.

8. *Advocacy*

In this attitude, the person goes beyond Nonviolence to actively seek ways to prevent the harm from ever occurring again and to dismantle structural violence. This attitude is similar to what John Wilmerding has described as "active peace," which has three components: peacemaking, peacekeeping, and peacebuilding.[22]

An important component of moving towards a harmless worldview is having alternate models available. The Institute of Noetic Sciences (IONS), for example, has a program of research on Worldview Transformation. It includes work on a Science of Peace and a curriculum for grade 8-12 called "Worldview Literacy" to help students increase their capacity for self-understanding and compassion.[23] For elementary school age children, there is now a video game called "Cool School: Where Peace Rules" that uses 52 different scenarios to teach peaceful ways to resolve conflicts.[24]

Gender Harmlessness as a Litmus Test

There are many contexts in which we harm ourselves and others. The most insidious, though, and the greatest human rights violation worldwide is in our attitudes and behavior towards women—gender violence. This includes the manifestations of self-hatred, prejudice, and violence towards women by women themselves. The hourly statistics are truly horrific,[25] and the UN Secretary-General has called repeatedly on the human family to end violence against women.[26]

While there are groups that are singled out for violent and abusive treatment for reasons other than gender, gender-based violence has several distinguishing features. Women and girls comprise approximately half of the world's population, making them the largest single group dealing with violence and discrimination. Also, while other targeted groups—ethnic or religious, for example—may live together in communities where they provide support for each other, women and girls usually live with the males who are most likely to be their abusers.[27]

Below are examples of how we might recognize and measure the diminution of three of the most damaging attitudes on the Harmlessness Scale™. These are the types of behavior that we must be able to bring to an end or actively discourage in order to be able to say that we have established an ethic of harmlessness.

Brutality

Despite international focus on violence against women, the most glaring types of brutality are still visited on hundreds of women and girls each minute. On the occasion of the 2010 celebration of International Women's Day, the UN High Commissioner for Human Rights, Navi Pillay, made the following observations:

It has been estimated that as many as one in three women across the world has been beaten, raped, or otherwise [physically] abused during the course of her lifetime. And the most common source of such violence comes from within the family. Amongst the most extreme forms of abuse is what is known as "honor killing." . . . In the name of preserving family "honor," women and girls are shot, stoned, burned, buried alive, strangled, smothered, and knifed to death with horrifying regularity. . . . The problem is exacerbated by the fact that in a number of countries domestic legal systems . . . still fully or partially exempt individuals guilty of honor killings from punishment. Perpetrators may even be treated with admiration and given special status within their communities.

Honor killings are, however, not something that can be simply brushed aside as some bizarre and retrograde atrocity that happens somewhere else. They are an extreme symptom of discrimination against women, which—including other forms of domestic violence—is a plague that affects every country. . . . For many women and girls, the family life that is supposed to be productive, protective, and harmonious is little more than a myth. Instead, for such females, family life means physical, sexual, emotional, or economic violence at the hands of an intimate partner or other family members. Domestic violence typically involves punches, kicks and slaps, or assaults with objects or weapons. It also frequently involves persistent belittlement and humiliation, and often includes the isolation of women from traditional supporters such as other family members and friends. Sometimes it may involve forced participation in degrading sexual acts, rape and homicide. . . . The reality for most victims, including victims of honor killings, is that state institutions fail them and that most perpetrators of domestic violence can rely on a culture of impunity for the acts they commit—acts that would often be considered as crimes, and be punished as such, if they were committed against strangers.

Traditionally, there has been some debate around the issue of state responsibility for acts committed in the private sphere. Some have argued, and continue to argue, that family violence is placed outside the conceptual framework of international human rights. However, under international laws and standards, there is a clear State responsibility to uphold women's rights and ensure freedom from discrimination, which includes the responsibility to prevent, protect, and provide redress—regardless of sex, and regardless of a person's status in the family.[28]

Statistics collected by the UN Development Fund for Women (UNIFEM) place the percentage of women worldwide who have experienced gender-based physical violence even higher—up to at least 70 percent.[29] And these statistics do not necessarily include a range of forms of ritual mutilation still performed regularly on females, usually by family members and under the most primitive and unhygienic of circumstances—for example, clitoral mutilation, infibulation, or breast ironing.

Statistics collected by national law enforcement agencies, as well as by agencies like UNIFEM and the World Health Organization are the easiest to monitor.[30] When the incidence of violence drops to close to zero, we will know we have made significant progress towards embedding harmlessness in our societal consciousness.

Harassment

Inciting hatred—and possible physical violence—against a specific group is an extreme form of harassment, generally punishable as a hate crime. In a bizarre legal loophole, inciting hatred against women is not illegal in Canada, according to the March 2010 ruling of a Quebec judge.[31] Defendant Jean-Claude Rochefort had created a website dedicated to Marc Lépine who murdered 14 women at Montreal's Ecole Polytechnique in 1989, calling him "Saint Marc" and wondering when it "could happen again, with the right people and the right equipment." Because women are not a minority nor are they named by Parliament as a group to be protected from the incitement of hatred, there is no crime in the Criminal Code of inciting hatred specifically against women.

In another type of harassment, controversy has been raging in countries where marriage of prepubescent females is seen as a legitimate way for fathers to settle debts or earn money. Publicity around a nine-year old Yemini girl who successfully filed for divorce started investigations, which resulted in a 2009 report from the Yemini Ministry of Social Affairs that a quarter of all females in Yemen marry before the age of 15.[32]

While statistics on all forms of harassment may not yet be readily available, we can monitor the decrease of this form of gender violence. One avenue of international monitoring is by tracking the implementation of the 1979 Convention on the Elimination of All Forms of Discrimination Against Women. Oversight of Member State compliance is provided by the Committee on the Elimination of Discrimination Against Women (CEDAW), which is one of the United Nation's eight human rights treaty bodies.[33]

Dismissiveness

The author Dorothy Sayers, in her essays on *Are Women Human?*, provides us with a useful and humorous perspective on the consequences of viewing persons entirely based on a single attribute, such as gender:

Probably no man has ever troubled to imagine how strange his life would appear to himself if it were unrelentingly assessed in terms of his maleness; if everything he wore, said, or did had to be justified by reference to female approval. . . If he were vexed by continual advice . . . how to be learned without losing his masculine appeal . . . In any book on sociology he would find, after the main portion dealing with human needs and rights, a supplementary chapter devoted to "The Position of the Male in the Perfect State." His newspaper would assist him with a "Men's Corner," telling him how, by the expenditure of a good deal of money and a couple of hours a day, he could attract the girls and retain his wife's affection; and when he had succeeded in capturing a mate, his name would be taken from him, and society would present him with a special title to proclaim his achievement. People would write books called, "History of the Male,". . . or "Psychology of the Male," and he would be regaled daily with headlines, such as "Gentleman-Doctor's Discovery." . . . He would be edified by . . . cross-shots of public affairs "from the masculine angle". . . . If, after a few centuries of this kind of treatment, the male was a little self-conscious, a little on the defensive, and a little bewildered about what was required of him, I should not blame him. . . . It would be more surprising if he retained any rag of sanity and self-respect.[34]

Here we begin to get into behaviors that are even harder to monitor. One way that we can track improvement in this dynamic is through statistics reported annually in the United Nations Development Programme (UNDP)'s *Human Development Report*.[35] In 1995, the *Human Development Report* began paying particular attention to gender and development and now includes two measures that are relevant to gender discrimination. The Gender-Related Development Index (GDI) looks at the degree of gender discrepancy on measures like life expectancy, education, and standard of living. The Gender Empowerment Measure (GEM) looks at gender inequalities in political participation, economic participation, and control of economic resources. Another practical measure is tracking the percentage of multiple-author research articles with women scientists listed as the lead authors, which is currently at less than 20 percent.[36]

Gender-based violence, especially of the "dismissiveness" and "harassment" variation, is so wide spread that we almost don't notice it. So, to see it shrinking is an excellent measure of the progress we are making as a society towards an ethic of harmlessness.

Advocacy Examples

The issues of gender violence and discrimination can still seem overwhelming in their damaging effects and their ability to hold a pervasive ethic of harmfulness in place. Given the extent to which gender violence is still accepted as normal and even laudable behavior, a simplistic move to an ethic of reciprocity will not be enough. The dynamics underlying these tentacles of violence are too tenacious and widespread, damaging the recipient and the perpetrator. We need multiple types of advocacy to break the hold of this festering destructiveness that casts a blight on the human spirit.

Over the years, it has been the persistent efforts of women's grassroots organizations worldwide that have focused attention on the issue of violence against women. Now those groups are being joined by many additional advocates. Within the United Nations, there is already an International Day for the Eradication of Violence Against Women — November 25 — to focus world attention on the issue. The UN Secretary-General began the UNiTE campaign to End Violence Against

Women, in 2008, saying: "We must unite. Violence against women cannot be tolerated in any form, in any context, in any circumstance, by any political leader or by any government."[37] UNiTE brings together UN agencies and offices to prevent and punish violence against women. In November 2009, the Secretary-General added a Network of Men Leaders saying, "Men must teach each other that real men do not violate or oppress women—and that a woman's place is not just in the home or the field, but in schools and offices and boardrooms.[38]

In March 2010, UNIFEM launched the Global Virtual Knowledge Centre to End Violence Against Women and Girls. Its purpose is to make available "lessons learned to date and recommended practices gleaned from initiatives on ending violence against women and girls, whether originating from the women's movement, civil society organizations, governments, the United Nations system, or other actors."[39]

In the private sector, one of the largest movements of men addressing issues of violence against women is the White Ribbon Campaign, begun in Canada in 1991.[40] Its purpose is to engage men and boys in making the personal commitment to "never commit, condone, or remain silent about violence against women and girls." There are now men's White Ribbon groups in 55 countries engaged in working on ending violence against women in all its forms, particularly through the education of men and boys.

We are also seeing some significant changes in a part of the human family that has historically been most committed to gender apartheid and gender violence. Over the vocal objections of some religious leaders and the equally vocal support of others, King Abdullah bin Abdul Aziz of Saudi Arabia launched the first coeducational university in that Kingdom in September 2009—the King Abdullah University for Science and Technology (KAUST).[41] King Abdullah has signaled that outdated social strictures will need to change if the Kingdom is to develop into a modern, diversified, less oil-dependent economy.

If we think of gender violence as a thick sheet of ice immobilizing us from living joyous and productive lives together, we recognize that that sheet of ice cannot simply be wished away nor will a single action disperse it. But if we have many committed initiatives, the ice will begin to crack and eventually break into small enough pieces that they can dissolve away.

Practicing Harmlessness

We are community builders. Our basic nature, unless distorted by fear, is cooperative and altruistic. We have only to look at the generosity of people after recent disasters—the 2004 Tsunami, the 2005 Hurricane Katrina, the 2010 Haiti earthquake. If we can but feel it, harmlessness is at the core of our beings, our most joyous way of life.

Metaphysical teachings, as recorded by Alice Bailey, instruct us that it is up to us to change harmful conditions by:

> . . . the development in ourselves of harmlessness. Therefore, study yourself from this angle. Study your daily conduct and words and thoughts so as to make them utterly harmless. Set yourself to think those thoughts about yourself and others that will be constructive and positive and hence harmless in their effects. Study your emotional effect on others so that by no mood, no depression, and no emotional reaction can you harm another. . . . Practice harmlessness with zest and understanding. . . . Harmlessness is the expression of the life of individuals who realize themselves to be everywhere, who live consciously as a soul, whose nature is love, whose method is inclusiveness. . . . Harm-

lessness brings about in life caution in judgment, reticence in speech, ability to refrain from impulsive action, and the demonstration of a non-critical spirit. . . Let harmlessness, therefore, be the keynote of your life.[42]

As we work together to embed an ethic of harmlessness, we become aware that the absence of overt harm is not enough. It will not even be enough to treat others respectfully and compassionately. We also need to dig out the root causes that hold harm in place as a tolerable and even desirable option. We need to experience our fundamental connectedness (the Principle of Interconnectivity) rather than distancing ourselves from others out of fear. And we need to recognize that we are indeed all part of the One Life and that our world is *not* basically dualistic (the Principle of Nonduality).[43]

Positive (proactive) harmlessness requires of us strength of character and a determination of purpose. It compels us to challenge the status quo, to question the forms of harm that flow — generally unchallenged — through our everyday lives. It mandates that we find ways to end and heal violence without resorting to violence ourselves, maintaining compassion for both the recipient and the perpetrator. Harmlessness is not for the faint of heart or for those longing for social approval. It may be the unpopular stand, the nagging reminder that we can do better.

Now that we are aware of how intimately harm is woven into our lives, the challenge that confronts us is how to be together in an empowering and joyous way. We must imbue the practice of positive harmlessness with as much attractive dynamism as is currently associated with violence. Then we may truly experience the dynamic that Nobel Peace Laureate Dr. Shirin Abadi has described: "Peace begins from inside: it boils from within, spreads through the family, saturates the society, and then covers the international arena."[44]

APPENDIX A*
UN Universal Declaration of Human Rights

PREAMBLE

Whereas recognition of the inherent dignity and of the equal and inalienable rights of all members of the human family is the foundation of freedom, justice and peace in the world,

Whereas disregard and contempt for human rights have resulted in barbarous acts that have outraged the conscience of humanity, and the advent of a world in which human beings shall enjoy freedom of speech and belief and freedom from fear and want has been proclaimed as the highest aspiration of the common people,

Whereas it is essential, if people are not to be compelled to have recourse, as a last resort, to rebellion against tyranny and oppression, that human rights should be protected by the rule of law,

Whereas it is essential to promote the development of friendly relations between nations,

Whereas the peoples of the United Nations have in the Charter reaffirmed their faith in fundamental human rights, in the dignity and worth of the human person, and in the equal rights of men and women and have determined to promote social progress and better standards of life in larger freedom,

Whereas Member States have pledged themselves to achieve, in cooperation with the United Nations, the promotion of universal respect for and observance of human rights and fundamental freedoms,

Whereas a common understanding of these rights and freedoms is of the greatest importance for the full realization of this pledge,

Now, Therefore THE GENERAL ASSEMBLY proclaims **THIS UNIVERSAL DECLARATION OF HUMAN RIGHTS** as a common standard of achievement for all peoples and all nations, to the end that every individual and every organ of society, keeping this Declaration constantly in mind, shall strive by teaching and education to promote respect for these rights and freedoms and by progressive measures, national and international, to secure their universal and effective recognition and observance, both among the peoples of Member States themselves and among the peoples of territories under their jurisdiction.

Article 1.
All human beings are born free and equal in dignity and rights. They are endowed with reason and conscience and should act towards one another in a spirit of cooperation.

Article 2.
Everyone is entitled to all the rights and freedoms set forth in this Declaration, without distinction of any kind, such as race, color, sex, language, religion, political or other opinion, national or social origin, property, birth, or other status. Furthermore, no distinction shall be made on the basis of the political, jurisdictional, or international status of the country or territory to which a person belongs, whether it be independent, trust, non-self-governing, or under any other limitation of sovereignty.

Article 3.
Everyone has the right to life, liberty, and security of person.

Article 4.
No one shall be held in slavery or servitude; slavery and the slave trade shall be prohibited in all their forms.

Article 5.
No one shall be subjected to torture or to cruel, inhuman, or degrading treatment or punishment.

Article 6.
Everyone has the right to recognition everywhere as a person before the law.

Article 7.
All are equal before the law and are entitled without any discrimination to equal protection of the law. All are entitled to equal protection against any discrimination in violation of this Declaration and against any incitement to such discrimination.

Article 8.
Everyone has the right to an effective remedy by the competent national tribunals for acts violating the fundamental rights granted them by the constitution or by law.

Article 9.
No one shall be subjected to arbitrary arrest, detention, or exile.

Article 10.
Everyone is entitled in full equality to a fair and public hearing by an independent and impartial tribunal, in the determination of their rights and obligations and of any criminal charge against them.

Article 11.
(1) Everyone charged with a penal offence has the right to be presumed innocent until proved guilty according to law in a public trial at which they have had all the guarantees necessary for their defense.

(2) No one shall be held guilty of any penal offence on account of any act or omission that did not constitute a penal offence, under national or international law, at the time when it was committed. Nor shall a heavier penalty be imposed than the one that was applicable at the time the penal offence was committed.

Article 12.
No one shall be subjected to arbitrary interference with their privacy, family, home, or correspondence, nor to attacks upon their honour and reputation. Everyone has the right to the protection of the law against such interference or attacks.

Article 13.
(1) Everyone has the right to freedom of movement and residence within the borders of each state.

(2) Everyone has the right to leave any country, including their own, and to return to their country.

Article 14.
(1) Everyone has the right to seek and to enjoy in other countries asylum from persecution.

(2) This right may not be invoked in the case of prosecutions genuinely arising from non-political crimes or from acts contrary to the purposes and principles of the United Nations.

Article 15.
(1) Everyone has the right to a nationality.

(2) No one shall be arbitrarily deprived of their nationality nor denied the right to change their nationality.

Article 16.
(1) Men and women of full age, without any limitation due to race, nationality, or religion, have the right to marry and to found a family. They are entitled to equal rights as to marriage, during marriage, and at its dissolution.

(2) Marriage shall be entered into only with the free and full consent of the intending spouses.

(3) The family is the natural and fundamental group unit of society and is entitled to protection by society and the State.

Article 17.
(1) Everyone has the right to own property alone as well as in association with others.

(2) No one shall be arbitrarily deprived of their property.

Article 18.
Everyone has the right to freedom of thought, conscience, and religion; this right includes freedom to change their religion or belief, and freedom, either alone or in community with others and in public or private, to manifest their religion or belief in teaching, practice, worship, and observance.

Article 19.
Everyone has the right to freedom of opinion and expression; this right includes freedom to hold opinions without interference and to seek, receive, and impart information and ideas through any media and regardless of frontiers.

Article 20.
(1) Everyone has the right to freedom of peaceful assembly and association.

(2) No one may be compelled to belong to an association.

Article 21.
(1) Everyone has the right to take part in the government of their country, directly or through freely chosen representatives.

(2) Everyone has the right of equal access to public service in their country.

(3) The will of the people shall be the basis of the authority of government; this will shall be expressed in periodic and genuine elections that shall be by universal and equal suffrage and shall be held by secret vote or by equivalent free voting procedures.

Article 22.
Everyone, as a member of society, has the right to social security and is entitled to realization, through national effort and international cooperation and in accordance with the organization and resources of each State, of the economic, social, and cultural rights indispensable for their dignity and the free development of their personality.

Article 23.
(1) Everyone has the right to work, to free choice of employment, to just and

favorable conditions of work and to protection against unemployment.

(2) Everyone, without any discrimination, has the right to equal pay for equal work.

(3) Everyone who works has the right to just and favorable remuneration ensuring for themselves and their families an existence worthy of human dignity, and supplemented, if necessary, by other means of social protection.

(4) Everyone has the right to form and to join trade unions for the protection of their interests.

Article 24.
Everyone has the right to rest and leisure, including reasonable limitation of working hours and periodic holidays with pay.

Article 25.
(1) Everyone has the right to a standard of living adequate for the health and well-being of themselves and of their families, including food, clothing, housing and medical care and necessary social services, and the right to security in the event of unemployment, sickness, disability, widowhood, old age, or other lack of livelihood in circumstances beyond their control.

(2) Motherhood and childhood are entitled to special care and assistance. All children, whether born in or out of wedlock, shall enjoy the same social protection.

Article 26.
(1) Everyone has the right to education. Education shall be free, at least in the elementary and fundamental stages. Elementary education shall be compulsory. Technical and professional education shall be made generally available and higher education shall be equally accessible to all on the basis of merit.

(2) Education shall be directed to the full development of the human personality and to the strengthening of respect for human rights and fundamental freedoms. It shall promote understanding, tolerance, and friendship among all nations, racial or religious groups, and shall further the activities of the United Nations for the maintenance of peace.

(3) Parents have a prior right to choose the kind of education that shall be given to their children.

Article 27.
(1) Everyone has the right freely to participate in the cultural life of the community, to enjoy the arts, and to share in scientific advancement and its benefits.

(2) Everyone has the right to the protection of the moral and material interests resulting from any scientific, literary, or artistic production of which they are the author.

Article 28.
Everyone is entitled to a social and international order in which the rights and freedoms set forth in this Declaration can be fully realized.

Article 29.
(1) Everyone has duties to the community in which alone the free and full development of their personality is possible.

(2) In the exercise of their rights and freedoms, everyone shall be subject only to such limitations as are determined by law solely for the purpose of securing due recognition and respect for the rights and freedoms of others and of meeting the just requirements of morality, public order, and the general welfare in a democratic society.

(3) These rights and freedoms may in no case be exercised contrary to the purposes and principles of the United Nations.

Article 30.
Nothing in this Declaration may be interpreted as implying for any State, group, or person any right to engage in any activity or to perform any act aimed at the destruction of any of the rights and freedoms set forth herein.

*Source: http://www.un.org/en/documents/udhr/. Please note that the text has been edited to ensure gender neutral language and U.S. spelling in keeping with the rest of this book.

APPENDIX B

Excerpts from the Executive Summary* of the
Study of the United Nations Secretary-General

Ending Violence Against Women: From Words to Action

October 9, 2006

Violence against women is a form of discrimination and a violation of human rights. It causes untold misery, cutting short lives and leaving countless women living in pain and fear in every country in the world. It harms families across the generations, impoverishes communities, and reinforces other forms of violence throughout societies. Violence against women stops them from fulfilling their potential, restricts economic growth, and undermines development. The scope and extent of violence against women are a reflection of the degree and persistence of discrimination that women continue to face. It can only be eliminated, therefore, by addressing discrimination, promoting women's equality and empowerment, and ensuring that women's human rights are fulfilled.

All of humanity would benefit from an end to this violence, and there has been considerable progress in creating the international framework for achieving this. However, new forms of violence have emerged and, in some countries, advances towards equality and freedom from violence previously made by women have been eroded or are under threat. The continued prevalence of violence against women is testimony to the fact that States have yet to tackle it with the necessary political commitment, visibility, and resources.

Violence against women is neither unchanging nor inevitable and could be radically reduced, and eventually eliminated, with the necessary political will and resources....

Overview

Violence against women was drawn out of the private domain into public attention and the arena of State accountability largely because of the grass-roots work of women's organizations and movements around the world. This work drew attention to the fact that violence against women is not the result of random, individual acts of misconduct, but rather is deeply rooted in structural relationships of inequality between women and men. The interaction between women's advocacy and United Nations initiatives has been a driving factor in establishing violence against women as a human rights issue on the international agenda.

There has been significant progress in elaborating and agreeing on international standards and norms. International and regional legal and policy instruments have clarified the obligations on States to prevent, eradicate, and punish violence against women. However, States around the world are failing to meet the requirements of the international legal and policy framework.

Causes and Risk Factors

The roots of violence against women lie in historically unequal power relations between men and women and pervasive discrimination against women in both the public and private spheres. Patriarchal disparities of power, discriminatory cultural norms, and economic inequalities serve to deny women's human rights and perpetuate violence. Violence against women is one of the key means through which male control over women's agency and sexuality is maintained. . . .

Violence against women is not confined to a specific culture, region, or country, or to particular groups of women within a society. The different manifestations of such violence and women's personal experiences are, however, shaped by factors such as ethnicity, class, age, sexual orientation, disability, nationality, and religion.

Forms and Consequences

There are many different forms of violence against women—physical, sexual, psychological, and economic. Some increase in importance while others diminish as societies undergo demographic changes, economic restructuring, and social and cultural shifts. For example, new technologies may generate new forms of violence, such as internet or mobile telephone stalking. Some forms, such as international [sex] trafficking and violence against migrant workers, cross national boundaries.

Women are subjected to violence in a wide range of settings, including the family, the community, state custody, and armed conflict and its aftermath. Violence constitutes a continuum across the lifespan of women, from before birth to old age. It cuts across both the public and the private spheres.

The most common form of violence experienced by women globally is intimate partner violence, sometimes leading to death. Also widespread are harmful traditional practices, including early and forced marriage and female genital mutilation/cutting. Within the community setting, femicide (gender-based murder of women), sexual violence, sexual harassment, and trafficking in women are receiving increasing attention.

Violence perpetrated by the State, through its agents, through omission, or through public policy, spans physical, sexual, and psychological violence. It can constitute torture. The high incidence of violence against women in armed conflict, particularly sexual violence including rape, has become progressively clearer.

Violence against women has far-reaching consequences for women, their children, and society as a whole. Women who experience violence suffer a range of health problems, and their ability to earn a living and to participate in public life is diminished. Their children are significantly more at risk of health problems, poor school performance, and behavioral disturbances.

Violence against women impoverishes women, their families, communities, and nations. It lowers economic production, drains resources from public services and employers, and reduces human capital formation. While even the most comprehensive surveys to date underestimate the costs, they all show that the failure to address violence against women has serious economic consequences.

There is compelling evidence that violence against women is severe and pervasive throughout the world...

State Responsibility

States have concrete and clear obligations to address violence against women, whether committed by state agents or by non-state actors. States are accountable to women themselves, to all their citizens, and to the international community. States have a duty to prevent acts of violence against women; to investigate such acts when they occur and prosecute and punish perpetrators; and to provide redress and relief to the victims.

While differing circumstances and constraints require different types of action to be taken by the State, they do not excuse State inaction. Yet States worldwide are failing to implement in full the international standards on violence against women.

When the State fails to hold the perpetrators of violence accountable, this not only encourages further abuses, it also gives the message that male violence against women is acceptable or normal. The result of such impunity is not only denial of justice to the individual victims/survivors, but also reinforcement of prevailing inequalities that affect other women and girls as well. . . .

*For the complete Executive Summary, see http://www.un.org/womenwatch/daw/vaw/launch/english/v.a.w-exeE-use.pdf.

APPENDIX C

Harmlessness Scale™: Questions & Scoring

Use of the Harmlessness Scale™ Questions

The Harmlessness Scale™ and the Harmlessness Scale™ Questions are the copyright of the author, Dorothy I. Riddle. You are welcome to use the Harmlessness Scale™ Questions for your individual personal use. There are three ways in which you might use the Harmlessness Scale™ Questions:

1. Before reading any explanation of the Harmlessness Scale™, you could answer all 20 questions by choosing the option that best represents how you think you would respond. Then, using the Harmlessness Scale™ Scoring Matrix, you could determine:

    o How likely you are to respond harmfully out of habit (A, with 40 = highest).
    o How likely you are to respond harmlessly out of habit (B, with 40 = highest).
    o Your likelihood, on balance, of behaving harmlessly (C, with +40 = highest).
    o The Attitude that is most habitual for you (D).

2. After you have read an explanation of the Harmlessness Scale™, you could answer the 20 questions in the way that you believe you would respond, not according to what response you feel would be the best. This is challenging to do! You would then be able to calculate the same four scores as listed above.

3. You could choose instead to use the Harmlessness Scale™ Questions to learn the distinctions between the eight attitudes. In this case, you would identify the attitude represented by each response option and then compare your answers to the Harmlessness Scale™ Scoring Matrix.

For those wishing to use the Harmlessness Scale™ Questions with groups for either research or training purposes, the author grants you permission to reproduce the Harmlessness Scale™ and the Harmlessness Scale™ Questions with the following stipulations:

- Clearly note on any reproduction: "© 2010 Dorothy I. Riddle. All rights reserved."

- Credit the Harmlessness Scale™ and the Harmlessness Scale™ Questions to Dorothy I. Riddle, *Positive Harmlessness in Practice* (Bloomington, IN: AuthorHouse, 2010, 235-243).

- Report the details of its use and any research results to *support@EnoughForUsAll.com*.

Harmlessness Scale™ Questions

> Circle the option for each question that best reflects your most likely response (even if you don't care for any of them), **not** what you think is the "best" response.

1. Your niece announces that she is pregnant and plans to keep working until two weeks before her due date. She's not sure how soon she'll go back to work after the birth. You say:
    a. "How can you even consider going back to work? Don't you want to be a good mother and stay home for the first five years?"
    b. "Just follow your instincts. I'm sure you'll make the right decision."
    c. To her partner, "What do you think is best?"
    d. To her partner, "How much parental leave are you planning to take?"

2. You are at lunch and someone makes a joke about dumb blondes. You:
    a. Laugh and say "Ain't that the truth!"
    b. Smile and say, "What can you expect?"
    c. Blurt out, "Are you an idiot? That's not funny.".
    d. Don't smile and remain silent.

3. You go to see a well-acted play at the community theatre. Part of the story line involves a man slapping his wife frequently in a fumbling attempt to discipline her. After the play, you comment:
    a. "It was really well acted."
    b. "I wish there hadn't been so much slapping but it really moved the plot line along."
    c. "I'm going to write to the Director of the community theater. Surely we can find good plays that don't model violent behavior."
    d. "I couldn't bear to stay and watch and had to leave."

4. You overhear two men talking about buying a five year old boy on the Internet for sexual play. You:
    a. Ignore them because they're probably joking.
    b. Tell them that you are appalled that they would consider doing that or even think it was funny.

c. Call the Cybercrime Unit to report them.
d. Comment to a friend that they sure have some creative ideas for fun.

5. Someone is taunting you about not having performed as well as they did. You:
    a. Ignore them and don't let their comments bother you.
    b. Yell at them that they didn't do so well themselves.
    c. Say nothing but feel terrible and like you are a failure.
    d. Use their comments as feedback to understand how you could improve.

6. You are at a restaurant and the man at the next table is yelling at his four-year-old to sit still. You:
    a. Keep quiet but smile at the child as you leave.
    b. Lean over and compliment the child on behaving so well in a public place.
    c. Ask the manager to get him to stop yelling because it is disturbing you.
    d. Yell at the man, "Stop it! How do you like being yelled at in public?"

7. You walk into a committee meeting just as one of two women on the committee is asked to take the minutes for the second meeting in a row. You comment:
    a. "I'll do it. I haven't taken the minutes yet."
    b. "Surely we could rotate so that everyone takes a turn?"
    c. (When someone else objects) "Why all this fuss? Let's get on with the meeting."
    d. (When someone else objects) "I'm sure she doesn't mind. She did a great job last time."

8. Your niece has dreams of becoming an astronaut and is doing well in science and math. You:
    a. Congratulate her.
    b. Canvas friends of yours to see how to introduce her to someone in the aerospace program.
    c. Point out that there are many hurdles for women in the aerospace program and wonder if she might not be better off teaching science and math.
    d. Make no comment because you're sure she's unlikely to

follow through.

9. You have a report due tomorrow morning, with at least 10 hours of work remaining, and it is already 4 pm. You:
    a. Stay up all night fueling yourself with caffeine and sugar.
    b. Negotiate to submit the report by the end of tomorrow, giving yourself more time so that you can get some sleep.
    c. Restructure the report so that you can get it done in five hours.
    d. Get upset with yourself for not having it done already.

10. You have just finished placing an order in a hardware store for bathroom fixtures and are told "the girls will call you when they're in." You:
    a. Thank them and leave without a comment.
    b. Ask them, "Surely you didn't mean to call your excellent support staff 'girls'?"
    c. Ask if the older woman behind the customer service counter is the one who will call.
    d. Yell, "What a sexist thing to say!"

11. You are on a conference planning committee and notice that all 15 proposed speakers are male. You say:
    a. Nothing as they are all well-known speakers.
    b. "Aren't there any women that we could invite to speak?"
    c. "I'd like to suggest a more balanced program. Here's a list of ten well-known women speakers to consider as well."
    d. Nothing, but later give the committee chair the list of women speakers.

12. Your young nephew is an excellent skater. You have just found out that he is excited about becoming an ice dancer after watching the Winter Olympics. You say to your nephew:
    a. "Are you nuts? What about hockey? That's a real man's sport."
    b. "Boy, your friends are sure going to laugh at you."
    c. "That's nice. Sounds like fun."
    d. "Hey, let's see if we can't get an ice dancing pair to come to your school for Career Day."

240   Enough for Us All: Positive Harmlessness

13. You are on the volunteer crew for the Writers' Festival and support tasks are being assigned. You notice that the heavy lifting is being assigned to men, with all of the other support tasks being assigned to women. You say:
    a. "How about if we mix things up a bit? Roberta is used to hauling furniture around, and Jonathan did a great job with registration last year."
    b. Nothing because you don't want to embarrass the head of the volunteer crew.
    c. Nothing because you don't want to upset the meeting.
    d. (When someone objects) "Hey, what's your problem?"

14. You have just completed a complex negotiation and are being criticized publicly for how long it took. You:
    a. Write a Letter to the Editor outlining the negotiation process and thanking all those who made it successful.
    b. Comment to any that criticize you, "I'd like to see you do any better."
    c. Feel quietly proud and ignore the comments.
    d. Say nothing, but feel they are probably right.

15. The radio station has just announced that a young cougar has been sighted in the woods on the edge of town. You:
    a. Call a couple of friends and head out with rifles to kill it before it can hurt anyone.
    b. Call the Wildlife Agency to see if they can relocate it before it can hurt anyone.
    c. Comment appreciatively that others have gone out and killed it "for the good of the community."
    d. Suggest that the woods be posted with "cougar sighted" so that people don't get hurt before there is a more permanent solution.

16. Your Town Council is about to vote to spend funds on setting up a wind power farm in a field at the edge of town. You:
    a. Prepare and present a report on the negative impact of wind farms on humans and bird life, including results of studies showing that wind farms are not environmentally cost effective.
    b. Send letters to the various Council members telling them that they are idiots for even considering this project.

c. Write a Letter to the Editor in support of this excellent environmental initiative and pointing out that the loss of some birds is a small price to pay for a cheap source of energy.
d. Meet privately with each Council member to see if you can persuade them that this is not a good plan.

17. While at a shopping mall, you see a woman slapping her young child and yelling, "Don't you talk back to me!" You:
    a. Step up to the woman and suggest that there might be a better way to handle matters.
    b. Quickly step up to her and slap her, saying, "Hey, stop it! How do you like being treated the way you're treating your child?"
    c. Say, "Boy, you are sure a poor excuse for a mother."
    d. Do nothing as the child is not yours.

18. You and a colleague are having a business lunch in a restaurant known for rapid service; however, you have been waiting 45 minutes and still haven't been served. You:
    a. Grab your waiter and say, "Can't you do your job properly and get us served?"
    b. Wait quietly, sure that there must be a crisis in the kitchen.
    c. Ask your waiter politely how much longer it will be.
    d. Get up and speak quietly to the person in charge, asking if you could get served as quickly as possible as you have been waiting.

19. You are waiting to board a flight whose aircraft was delayed inbound, and you notice that the gate security screener is pulling aside only passengers with dark hair and dark complexioned skin. You say:
    a. Nothing, as they have not pulled you aside.
    b. (When you hear a complaint) "Hey, they are just doing their job."
    c. (When you hear a complaint) "Hey, keep quiet. Do you want to delay us further?"
    d. (To the airline check-in staff) "Could you give me contact information so that I can write to airport security? This passenger selection process doesn't seem fair."

20. You turn the corner on a quiet street and find three young men who are tormenting a shaky elderly woman by jumping in front of her when she tries to walk forward. You:
    a. Go get a police officer to intervene.
    b. Turn back quickly so that you are not noticed.
    c. Go get a police officer to intervene, and then make a proposal to the Town Council that a system be set up where vulnerable persons can get escorts to keep them safe on the streets as is done on some college campuses.
    d. Walk quickly by, thinking she asked for it by being alone.

© 2010 Dorothy I. Riddle. All rights reserved.

Scoring Instructions

1. Once you have circled your *most likely* response for each question, then turn to the Scoring Matrix on the next page and circle your 20 responses.

2. Add up the number of responses you have circled in each column and write that number in the "Number Circled" row. The numbers in that row should total 20.

3. Then multiply each of the "Number Circled" values by the number given below it in the "Multiply by" row, and write the result in the "Total" row.

4. Calculate your values for "A," "B," and "C."

5. Then select as "D" the Attitude for which you have the highest value in the "Number Circled" row:

    1 = Brutality
    2 = Harassment
    3 = Dismissiveness
    4 = Defensiveness
    5 = Abstention
    6 = Supportiveness
    7 = Nonviolence
    8 = Advocacy

Harmlessness Scale™ Scoring Matrix

Question Number	Attitude							
	1	2	3	4	5	6	7	8
1	a		c			b	d	
2	c	a		b	d			
3			a	b	d			c
4			a	d		b	c	
5	b			c	a	d		
6	d		c		a	b		
7			c	d	a			b
8		c	d			a		b
9	a	d				c	b	
10	d			a		c	b	
11			a			d	b	c
12	a	b			c			d
13		d		c	b		a	
14		b		d	c			a
15	a			c			b	d
16	c	b				d		a
17	b	c	d				a	
18		a			b	d	c	
19		c	a		b			d
20			b	d			a	c
Number Circled	___	___	___	___	___	___	___	___
Multiply by	x4 =	x3 =	x2 =	x1 =	x1 =	x2 =	x3 =	x4 =
Total	___	___	___	___	___	___	___	___

Scoring		
	A: ___ Add totals for columns 1-4	**B:** ___ Add totals for columns 5-8
	C: ___ Subtract A from B	**D:** Attitude with highest number on "Number Circled" row is ___.

Key:
A: How likely you are to respond harmfully out of habit (40 = highest).
B: How likely you are to respond harmlessly out of habit (40 = highest).
C: Your likelihood, on balance, of behaving harmlessly (+40 = highest).
D: The Attitude that is most habitual for you.

Notes

Chapter 1: The Concept of Harm

1. Ronald Grisanti, "Iatrogenic Disease: The 3rd Most Fatal Disease in the USA," http://www.yourmedicaldetective.com/public/335.cfm.

2. See http://www.who.int/violence_injury_prevention/publications/violence/explaining/en/index.html.

3. The Canadian Media Awareness Network provides summaries of current issues at http://www.media-awareness.ca/english/issues/violence/violence_entertainment.cfm, including a range of specific examples and a summary of research at http://www.media-wareness.ca/english/issues/violence/effects_media_violence.cfm. Examples of other sites summarizing findings on violence in the media include http://www.lion-lamb.org/media_violence.htm and http://www.parentstv.org/ptc/facts/mediafacts.asp. The growing concern about violence in music has been researched by Craig A. Anderson and Nicholas L. Carnagey, "Exposure to Violent Media: The Effects of Songs With Violent Lyrics on Aggressive Thoughts and Feelings," *Journal of Personality and Social Psychology* 84, no. 5 (2003): 960–971.

4. Published in *Pediatrics*, October 19, 2009.

5. See statistics at http://www.who.int/mediacentre/factsheets/fs310/en/index.html and http://www.who.int/features/factfiles/injuries/en/index.html.

6. See http://www.who.int/violence_injury_prevention/violence/global_campaign/en/.

7. See http://www.hrw.org/en/news/2006/10/08/un-new-report-says-violence-against-women-human-rights-violation.

8. See http://www.un.org/en/women/endviolence/.
9. See http://www.un.org/millenniumgoals/education.shtml.
10. See http://www.uri.org/.
11. Stuart Berg Flexner and Leonore Crary Hanck, eds. *The Random House Dictionary of the English Language*, 2nd ed. (New York: Random House, 1987).
12. See http://www.armenian-genocide.org/us-5-29-15.html.
13. The Nuremberg trials were convened under the London Charter of the International Military Tribunal issued August 8, 1945, and were quickly followed by the Tokyo War Crimes Tribunal in 1946. After a hiatus, the International Criminal Tribunal for the former Yugoslavia was established by the UN Security Council in 1993, which then set up the International Criminal Tribunal for Rwanda in 1994 whose success in prosecuting genocide was viewed by UN Secretary-General Kofi Annan as "a historic milestone." Additional special international tribunals have included the Special Tribunal for Cambodia, the Special Court for Sierra Leone, and the Special Tribunal for Lebanon.
14. The Geneva Conventions, plus the Additional Protocols (1977 and 2005), form the basis of international humanitarian law to limit the barbarity of war.
15. The full text is reproduced in Appendix A, from http://www.un.org/en/documents/udhr/.
16. The various conventions used to elaborate and enforce the UN Universal Declaration of Human Rights are listed at http://www2.ohchr.org/english/law/. Some of the specific conventions already in force are:
    o The Convention on the Prevention and Punishment of the Crime of Genocide (*entered into force* January 12, 1951).
    o The International Convention on the Elimination of All Forms of Racial Discrimination (*entered into force* January 4, 1969).
    o The International Covenant on Economic, Social, and Cultural Rights (*entered into force* January 3, 1976).

- The International Covenant on Civil and Political Rights (*entered into force* March 23, 1976).

- The Convention on the Elimination of All Forms of Discrimination Against Women (*entered into force* September 3, 1981).

- The Convention Against Torture and Other Cruel, Inhuman, or Degrading Treatment or Punishment (*entered into force* June 26, 1987).

- The Convention on the Rights of the Child (*entered into force* September 2, 1990).

17. The Statute of the International Criminal Court (ICC), known as the Rome Statute, was adopted in Rome in July 1998. Under it, the ICC has launched proceedings regarding crimes committed in the Central African Republic, Darfur, the Democratic Republic of the Congo, and northern Uganda.

18. Mark Juergensmeyer, *Terror in the Mind of God: The Global Rise of Religious Violence* (Berkeley: University of California Press, 2000), 149.

19. Recent examples of U.S. suicides as a result of bullying, which included cyber bullying (Facebook postings, instant messaging, text messaging), include Phoebe Prince (15) who killed herself January 14, 2010, and Alexi Pilkingon (14) who killed herself March 21, 2010. In both instances, not only were they tormented when alive, but nasty messages continued after their deaths on Facebook Memorial pages. Forty-two percent of U.S. teenagers report that they have been the targets of cyber bullying.

20. Michael Price, "Revenge and the People Who Seek It: New Research Offers Insights Into the Dish Best Served Cold," *Monitor on Psychology*, June 2009: 34-37.

21. See http://www1.jca.apc.org/vaww-net-japan/english/womenstribunal2000/oraljudgement.pdf.

22. Daniel B Wood, "How a Jury of Teens May Put a New Spin on Juvenile Justice," *The Christian Science Monitor* 102, no. 21 (April 1, 2010): 20-21. See also information on the National Association of Youth Courts at http://www.youthcourts.net/.

23. See http://scowinstitute.ca/library/documents/Aboriginal_Courts.pdf.

24. Azim Khamisa, *From Forgiveness to Fulfillment* (La Jolla, CA: ANK Publishing, 2007). See http://www.AzimKhamisa.com and the Tariq Khamisa Foundation at http://www.TKF.org.

Chapter 2: Positive Harmlessness as Our Core Value

1. Jean Pearsall, ed., *The New Oxford Dictionary of English* (Oxford: Clarendon Press, 1998).

2. Stuart Berg Flexner and Leonore Crary Hanck, eds. *The Random House Dictionary of the English Language*, 2nd ed. (New York: Random House, 1987).

3. Walter W. Avis, Patrick D. Drysdale, Robert J. Gregg et al., *Gage Canadian Dictionary* (Toronto: Gage Educational Publishers, 1983).

4. Alice A. Bailey, *The Light of the Soul* (New York: Lucis Publishing, 1927), 184-190.

5. Noam Chomsky, "Terror and Just Response," *ZNet*, July 2, 2002.

6. Alice A. Bailey, *A Treatise on White Magic* (New York: Lucis Publishing, 1934), 490.

7. Alice A. Bailey, *Esoteric Healing* (New York: Lucis Publishing, 1953), 670.

8. Dorothy I. Riddle, *Principles of Abundance for the Cosmic Citizen: Enough for Us All, Volume One* (Bloomington, IN: AuthorHouse, 2010).

9. Ibid., Chapter 5.

10. Susanne Cook-Greuter, "Mature Ego Development: A Gateway to Ego Transcendence?" *Journal of Adult Development* 7, no. 4 (October 2000), 234.

11. Sadie F. Dingfelder, "How Artists See," *Monitor on Psychology* 41, no. 2 (February 2010): 40.

12. Lynn Margulis, *The Symbiotic Planet: A New Look at Evolution* (London: Phoenix, 1998).

13. Sadie F. Dingfelder, "Nice by Nature?" *Monitor on Psychology* 58 (September 2009): 60-61.

14. Riddle, *Principles of Abundance*, Chapter 5.

15. Response given by the Dalai Lama during a question and answer session after his talk on "Cultivating Happiness" during the *Vancouver Dialogues*, which the author attended in Vancouver, BC, Canada on September 9, 2006. Session summary available at http://dalailamacenter.org/conference/session/cultivating-happiness.

16. In the Ageless Wisdom teachings, the One Life expresses Itself through seven Rays, or types of energy:
    Ray 1   Power, Will
    Ray 2   Love-Wisdom
    Ray 3   Active Creative Intelligence
    Ray 4   Harmony Through Conflict
    Ray 5   Concrete Knowledge
    Ray 6   Devotion, Idealism
    Ray 7   Organization, Order
    For a detailed discussion, see Alice A. Bailey, *Esoteric Psychology, Vol. I* (New York: Lucis Publishing, 1936).

17. See references at http://www.canada.com/story_print.html?id=53fc1139-9c97-44ab-9667-e6f70a5092fc&sponsor= and http://www.thehawnfoundation.org/wordpress/wp-content/uploads/2007/05/summary-of-the-effectiveness-of-the-me-program_april2009ksrfinal1.pdf.

Chapter 3: Harmlessness and the Butterfly Shift

1. "Gendercide: The Worldwide War on Baby Girls," *The Economist* 394, no. 8672 (March 6-12, 2010): 77-80.

2. Association for Psychological Science, "Bilingual Babies: The Roots of Bilingualism in Newborns," *ScienceDaily* (February 17, 2010), http://www.sciencedaily.com/releases/2010/02/100216142330.htm.

3. Society for Research in Child Development, "Awareness of Racism Affects How Children Do Socially and Academically,"

ScienceDaily (November 14, 2009), http://www.sciencedaily.com/releases/2009/11/091113083301.htm; University of Cincinnati, "Research Examines Stereotypes of Immigrants to the United States," *ScienceDaily* (August 10, 2009), http://www.sciencedaily.com/releases/2009/08/090810104807.htm.

4. University of Virginia, "Citizens in 34 Countries Show Implicit Bias Linking Males More Than Females With Science," *ScienceDaily* (June 23, 2009), http://www.sciencedaily.com/releases/2009/06/090622171410.htm.

5. Michael Price, "Making Sense of Dollars and Cents," *Monitor on Psychology* 39, no. 2 (February 2008): 34-36.

6. Jane Perlez, "Romanian 'Orphans': Prisoners of Their Cribs," *The New York Times*, March 25, 1996; Kate McGowen, "What Happened to Romania's Orphans?" *BBC News*, July 8, 2005.

7. See, for example, M. Fishbein, "An Investigation of the Relationships Between Beliefs About an Object and the Attitude Toward That Object," *Human Relations* 16 (1963): 233-239; M.J. Rosenberg, "Cognitive Structure and Attitudinal Affect," *Journal of Abnormal and Social Psychology* 53 (1956): 367-372; M.J. Rosenberg, "An Analysis of Affective-Cognitive Consistency," in *Attitude Organization and Change*, eds. M.J. Rosenberg, C.I. Hovland, W.J. McGuire, R.P. Abelson, and J.W. Brehm, 15-64 (New Haven: Yale University Press, 1960).

8. Association for Psychological Science, "Joint Attention Study Has Implications for Understanding Autism," *ScienceDaily* (September 29, 2007), http://www.sciencedaily.com/releases/2007/09/070926111521.htm; Wellcome Trust, "Babies' Brains Tuned to Sharing Attention With Others," *ScienceDaily* (January 27, 2010), http://www.sciencedaily.com/releases/2010/01/100126220331.htm.

9. Duke University Medical Center, "Neurons That 'Mirror' the Attention of Others Discovered," *ScienceDaily* (May 20, 2009), http://www.sciencedaily.com/2009/05/090518172451.htm.

10. David C. Korten, *Agenda for a New Economy: From Phantom Wealth to Real Wealth* (San Francisco: Berrett-Koehler Publishers, 2009), 184-185.

Chapter 4: Managing Our Focus

1. William James, *The Principles of Psychology, Vol. 1* (New York: Henry Holt, 1890), 403-404.

2. George A. Miller, "The Magical Number Seven, Plus or Minus Two: Some Limits on Our Capacity for Processing Information," *Psychological Review* 63 (1956): 81-97.

3. J.A. Easterbrook, "The Effect of Emotion on Cue Utilization and the Organization of Behavior," *Psychological Review* 66 (1959): 183-201.

4. Siri Carpenter, "Sights Unseen," *Monitor on Psychology* 32, no. 4 (April 4, 2001); M.M. Chun and R. Marois, "The Dark Side of Visual Attention," *Current Opinion in Neurobiology* 12, no. 2 (April 1, 2002): 184-189.

5. Ronald A. Rensink, J. Kevin O'Regan, and James J. Clark, "To See or Not to See: The Need for Attention to Perceive Changes in Scenes," *Psychological Science* 8, no. 5 (1997): 368-373; M. Silverman and A. Mack, "Change Blindness and Priming: When It Does and Does Not Occur," *Consciousness and Cognition* 15 (June 2006): 409-422.

6. Arien Mack and Irvin Rock, *Inattentional Blindness: An Overview* (Cambridge, MA: MIT Press, 1998).

7. David L. Strayer, Frank A. Drews, and William A. Johnston, "Cell Phone-Induced Failures of Visual Attention During Simulated Driving," *Journal of Experimental Psychology: Applied* 9, no. 1 (March 2003): 23-32.

8. David L. Strayer, Frank A. Drews, and Dennis J. Crouch, "A Comparison of the Cell Phone Driver and the Drunk Driver," *Human Factors* 48, no. 2 (Summer 2006): 381-391.

9. William Saletan, "The Mind-Blackberry Problem: Hey, You! Cell-Phone Zombie! Get Off the Road!" *Slate* (October 23, 2008), http://www.slate.com/id/2202978/.

10. Ellen J. Langer, *Mindfulness* (Reading, MA: Addison Wesley, 1989).

11. Isabel Briggs Myers and Peter B. Myers, *Gifts Differing: Understanding Personality Type* (Mountain View, CA: Davies-Black Publishing, 1980).

12. Karl Duncker, "On Problem Solving," *Psychological Monographs* 58, no. 5 (1945): Whole No. 270; R.E. Adamson, "Functional Fixedness as Related to Problem Solving: A Repetition of Three Experiments," *Journal of Experimental Psychology* 44 (1952): 288-291.

13. Daniel J. Siegel, *The Mindful Brain: Reflection and Attunement in the Cultivation of Well-Being* (New York: W.W. Norton, 2007).

14. See, for example, http://www.helpothers.org/index.php or http://www.actsofkindness.org/.

15. G.A. Bonanno, A. Papa, K. O'Neill, M. Westphal, and K. Coifman, "The Importance of Being Flexible: The Ability to Enhance and Suppress Emotional Expressions Predicts Long-Term Adjustment," *Psychological Science* 15 (2004): 482–487; J.M. Richards and J.J. Gross, "Composure at Any Cost? The Cognitive Consequences of Emotion Suppression," *Personality and Social Psychology Bulletin* 25 (1999): 1033–1044; J.M. Richards and J.J. Gross, "Emotion Regulation and Memory: The Cognitive Costs of Keeping One's Cool," *Journal of Personality and Social Psychology* 79 (2000): 410–424.

16. McKay Moore Sohlberg and Catherine A. Mateer, *Introduction to Cognitive Rehabilitation: Theory and Practice* (New York: Guilford Press, 1989).

17. See U.S. Better Business Bureau, "Complaint Handling—An Advantage for Businesses," http://us.smetoolkit.org/us/en/content/en/2855/Complaint-Handling-%E2%80%93-An-Advantage-for-Businesses. Also Frederic B. Kraft and Charles L. Martin, "Customer Compliments as More Than Complementary Feedback," *Journal of Consumer Satisfaction, Dissatisfaction and*

Complaining Behavior, 2001; Richard L. Oliver, *Satisfaction* (New York: McGraw-Hill, 1997).

Chapter 5: Noticing—Step One

1. Cell Press, "In 'Reading' a Gaze, What We Believe Changes What We See," *ScienceDaily* (June 26, 2009), http://www.sciencedaily.com/releases/2009/06/090625133049.htm.

Chapter 6: Leveraging Emotions

1. Alice A. Bailey, *Light of the Soul* (New York: Lucis Publishing, 1927), 20.

2. Candace B. Pert, *Molecules of Emotion: The Science Behind Mind-Body Medicine* (New York: Scribner, 1997), 312.

3. Brian Swimme, *The Hidden Heart of the Cosmos: Humanity and the New Story* (Maryknoll, NY: Orbis Books, 1996), 100. The term "all-nourishing abyss" contains a "dual emphasis: the universe's generative potentiality is indicated with the phrase 'all-nourishing,' but the universe's power of infinite absorption is indicated with 'abyss.'"

4. Robert C. Solomon, *The Passions: The Myth and Nature of Human Emotion* (New York: Anchor Press/Doubleday, 1976).

5. Institute of HeartMath, "The Power of Emotions," http://store.heartmath.org/store/solutions-for-stress/the-power-of-emotion, 4.

6. R.B. Zajonc, "Feelings and Thinking: Preferences Need No Inferences," *American Psychologist* 35, no. 2 (1980): 151-175.

7. Jonah Lehrer, *How We Decide* (New York: Houghton Mifflin Harcourt, 2009), xvi.

8. J. Campos, R.G. Campos, and K. Barrett, "Emergent Themes in the Study of Emotional Development and Emotion Regulation," *Developmental Psychology* 25, no. 3 (1989): 394-402; D. Keltner and J. Haidt, "Social Functions of Emotions at Four Levels of Analysis," *Cognition and Emotion* 13, no. 5 (1999): 505-521.

9. Shlomo Hareli and Anat Rafaeli, "Emotion Cycles: On the Social Influence of Emotion in Organizations," *Research in Organizational Behavior* 28 (2008): 35-59.

10. San Francisco University, "Facial Expressions of Emotion Are Innate, Not Learned," *ScienceDaily* (December 30, 2008), http://www.sciencedaily.com/releases/2008/12/081229080859.htm.

11. See Daniel Goleman, *Emotional Intelligence: Why It Can Matter More Than IQ* (New York: Bantam Books, 1995), though the concept was first presented in P. Salovey and J.D. Mayer, "Emotional Intelligence," *Imagination, Cognition, and Personality* 9 (1994): 185-211.

12. Katherine R. Mickley and Elizabeth A. Kensinger, "Emotional Valence Influences the Neural Correlates Associated with Remembering and Knowing," *Cognitive, Affective & Behavioral Neuroscience* 8, no. 2 (June 2008): 143-152.

13. Hong Li and Jianjin Yuan, "The Neural Mechanism Underlying the Female Advantage in Identifying Negative Emotions: An Event-Related Potential Study," *NeuroImage* 40, no. 4 (May 2008): 1921-1929.

14. Michael Edwardson, "Emotional Profiling," *Tempus*, Winter 1998: 5.

15. Barbara L. Fredrickson, "What Good Are Positive Emotions?" *Review of General Psychology* 2, no. 3 (1998): 300-319.

16. C.J. Brainerd, L.M. Stein, R.A. Silveria, G. Rohenkohl, and V.F. Reyna, "How Does Negative Emotion Cause False Memories?" *Psychological Science* 19, no. 9 (September 2008): 919-925.

17. E.A. Kensinger, "Remembering Emotional Experiences: The Contribution of Valence and Arousal," *Reviews in the Neurosciences* 15 (2004): 241-251; E.A. Kensinger and S. Corkin, "Memory Enhancement for Emotional Words: Are Emotional Words More Vividly Remembered Than Neutral Words?" *Memory and Cognition* 31 (2003): 1169-1180.

18. Jane Vincent, "Emotional Attachment and Mobile Phones," *Knowledge, Technology, & Policy* 19, no. 1 (Spring 2006): 39-44.

19. S.G. Barsade and D.E. Gibson, "Why Does Affect Matter in Organizations?" *Academy of Management Perspectives*, 2007: 35-59.

20. Anca M. Miron, Beverly Brummett, Brent Ruggles, and Jack W. Brehm, "Deterring Anger and Anger-Motivated Behaviors," *Basic & Applied Social Psychology* 30, no. 4 (October-December 2008): 326-338.

21. Robert Plutchik, "The Nature of Emotions," http://www.fractal.org/Bewustzijns-Besturings-Model/Nature-of-emotions.htm.

22. Wellcome Trust, "Everybody Laughs, Everybody Cries: Researchers Identify Universal Emotions," *ScienceDaily* (January 26, 2010), http://www.sciencedaily.com/releases/2010/01/100125173234.htm.

23. T. Sharot and E.A. Phelps, "How Arousal Modulates Memory: Disentangling the Effects of Attention and Retention," *Cognitive, Affective, & Behavioral Neuroscience* 4, no.3 (2004): 294-306.

24. K.N. Ochsner, "Are Affective Events Richly Recollected or Simply Familiar? The Experience and Process of Recognizing Feelings Past," *Journal of Experimental Psychology: General* 129 (2000): 242-261.

25. F. Heuer and D. Reisberg, "Vivid Memories of Emotional Events: The Accuracy of Remembered Minutiae," *Memory & Cognition* 18 (1990): 496–50; K.S. LaBar and E.A. Phelps, "Arousal-Mediated Memory Consolidation: Role of the Medial Temporal Lobe in Humans," *Psychological Science* 9 (1998): 490-493; T. Sharot and E.A. Phelps, "How Arousal Modulates Memory: Disentangling the Effects of Attention and Retention," *Cognitive, Affective, & Behavioral Neuroscience* 4, no. 3 (2004): 294-306.

26. Association for Research in Vision and Ophthalmology, "Are Angry Women More Like Men?" *ScienceDaily* (December 14, 2009), http://www.sciencedaily.com/releases/2009/12/091206110844.htm.

27. S.G. Barsade, "The Ripple Effect: Emotional Contagion and Its Influence on Group Behavior," *Administrative Science Quarterly*

47 (2002): 644-675; R. Neumann and F. Strack, "'Mood Contagion': The Automatic Transfer of Mood Between People," *Journal of Personality and Social Psychology* 79, no. 2 (2000): 211-223.

28. Joshua Freedman, *White Paper: Emotional Contagion*, April 2007, http://www.6seconds.org/modules.php?name=News&file=article&sid=267.

29. Albert Bandura, "Reflexive Empathy: On Predicting More Than Has Ever Been Observed," *Behavioral and Brain Sciences* 25 (2002): 24-25.

30. Dean Radin, "Intention and Reality: The Ghost in the Machine Returns," *Shift: At the Frontiers of Consciousness*, no. 15 (June-August 2007): 23-26.

31. Masaru Emoto, *The Hidden Messages in Water* (New York: Atria Books, 2001); Dean Radin, Gail Hayssen, Masaru Emoto, and Takashige Kizu, "Double-Blind Test of the Effects of Distant Intention on Water Crystal Formation," *Explore: The Journal of Science and Healing* 2, no. 5, (September 2006): 408–411.

32. Pert, *Molecules of Emotion*, 148.

33. Erasmus University Rotterdam, "Happiness Lengthens Life," *ScienceDaily* (August 5, 2008), http://www.sciencedaily.com/releases/2008/08/080805075614.htm; Institute of HeartMath, "Emotional Balance and Health," http://www.heartmath.org/research/science-of-the-heart-emotional-balance.html; R. Veenhoven, "Healthy Happiness: Effects of Happiness on Physical Health and the Consequences for Preventive Health Care," *Journal of Happiness Studies* 9 (2008): 449-469.

34. Stephanie L. Brown, R. Nesse, A.D. Vinokur, and D.M. Smith, "Providing Support May Be More Beneficial Than Receiving It: Results From a Prospective Study of Mortality," *Psychological Science* 14 (2003): 320-327.

35. Institute of HeartMath, "The Power of Emotions," http://store.heartmath.org/store/solutions-for-stress/the-power-of-emotion, 9.

36. Institute of HeartMath, "Head-Heart Interactions," http://www.heartmath.org/research/research-science-of-the-heart.html.

37. Ibid.

38. Jonah Lehrer, *How We Decide* (New York, Houghton Mifflin Harcourt, 2009).

39. Bhikkhu Bodhi, *The Noble Eightfold Path: Way to the End of Suffering* (Kandy, Sri Lanka: Buddhist Publication Society, 1994), 39.

40. See information about the TED prize at http://www.tedprize.org/.

41. See the Charter for Compassion at http://www.charterforcompassion.com/.

42. Institute of HeartMath, "Emotional Balance and Health," http://www.heartmath.org/research/science-of-the-heart-emotional-balance.html.

43. A.M. Wood, J. Maltby, N. Stewart, P.A. Linley, and S. Joseph, "A Social-Cognitive Model of Trait and State Levels of Gratitude," *Emotion* 8 (2008): 281-290.

44. Alex M. Wood, John Maltby, R. Gillett, P.A. Linley, and Stephen Joseph, "The Role of Gratitude in the Development of Social Support, Stress, and Depression: Two Longitudinal Studies," *Journal of Research in Personality* 42 (2008): 854-871.

45. M.E. McCullough, R.A. Emmons, and J. Tsang, "The Grateful Disposition: A Conceptual and Empirical Topography," *Journal of Personality and Social Psychology* 82 (2002): 112-127; Alex M. Wood, Stephan Joseph, and John Maltby, "Gratitude Uniquely Predicts Satisfaction with Life: Incremental Validity Above the Domains and Facets of the Five Factor Model," *Personality and Individual Differences* 45, no. 1 (2008): 49-54.

46. Alex M. Wood, Stephan Joseph, and John Maltby, "Gratitude Predicts Psychological Well-being above the Big Five Facets," *Personality and Individual Differences* 46, no. 4 (2009): 443-447.

47. Alex M. Wood, Stephen Joseph, and P.A. Linley, "Coping Style as a Psychological Resource of Grateful People," *Journal of Social and Clinical Psychology* 26 (2007): 1108-1125.

48. P.C. Watkins, J. Scheer, M. Ovnicek, and R. Kolts, "The Debt of Gratitude: Dissociating Gratitude and Indebtedness," *Cognition and Emotion* 20 (2006): 217-241.

49. J.A. Tsang, "The Effects of Helper Intention on Gratitude and Indebtedness," *Motivation and Emotion* 30 (2006): 199-205.

50. Alice A. Bailey, *Esoteric Psychology, Vol. I* (New York: Lucis Publishing, 1936), 49.

51. See details of Bhutan's Gross National Happiness index at http://www.grossnationalhappiness.com/.

52. Pert, *Molecules of Emotion*, 147.

Chapter 7: Feeling—Step Two

1. For more information, see http://www.laughteryoga.org/ and Ben Arnoldy, "Go Ahead and Laugh!" *The Christian Science Monitor* 102, no. 21 (April 2010): 4.

2. Malcolm Gladwell, *Blink: The Power of Thinking Without Thinking* (New York: Little, Brown and Company, 2005), 206.

3. This type of visualization is known as shielding. You can be as creative as you wish in creating a shield that works for you. For example, if you wish to protect yourself from others' negative feelings while having them experience the consequences of their emotional energy, you could visualize your bubble of light as lined with silver so that it becomes like a mirror reflecting their emotions back to them.

4. See http://www.gratefulness.org/.

5. Todd B. Kashdan, Anjali Mishra, William E. Breen, and Jeffrey J. Froh, "Gender Differences in Gratitude: Examining Appraisals, Narratives, the Willingness to Express Emotions, and Changes in Psychological Needs," *Journal of Personality* 77, no. 3 (June 2009): 691-730.

6. Eva Kaplan-Leiserson, "Put Your Heart Math Into It," http://www.heartmath.com/company/proom/articles/td_magazine.pdf

Chapter 8: Reviewing Our Action Options

1. A. Parasuraman, V.A. Zeithaml, and L.L. Berry, "A Conceptual Model of Service Quality and Its Implications for Future Research," *Journal of Marketing* 49 (Fall 1985): 41-50.

2. K.N. Ochsner, "Are Affective Events Richly Recollected or Simply Familiar? The Experience and Process of Recognizing Feelings Past," *Journal of Experimental Psychology: General* 129 (2000): 242-261; T. Sharot and E.A. Phelps, "How Arousal Modulates Memory: Disentangling the Effects of Attention and Retention," *Cognitive, Affective, & Behavioral Neuroscience* 4, no. 3 (2004): 294-306.

3. Abraham H. Maslow, "A Theory of Human Motivation," *Psychological Review* 50, no. 4 (1943): 370–396.

4. See the UN Universal Declaration of Human Rights in Appendix A.

5. The Sharing Circle, "The 7 Sacred Teachings," http://www.thesharingcircle.com/sacredteachings.html.

6. Cell Press, "Facial Expressions Show Language Barriers, Too," *ScienceDaily* (August 16, 2009), http://www.sciencedaily.com/releases/2009/08/090813142131.htm.

7. Edward T. Hall, *The Silent Language* (Greenwich, CT: Fawcett, 1959); *Beyond Culture* (Garden City, NJ: Anchor Press, 1976).

8. Florence R. Kluckhohn and Fred L. Strodtbeck, *Variations in Value Orientations* (Evanston, IL: Row, Peterson, 1961).

9. Geert Hofstede, *Culture's Consequences: International Differences in Work-Related Values* (Beverly Hills, CA: Sage Publications, 1984).

10. See suggestions at http://www.helpothers.org/ and http://www.actsofkindness.org/.

11. See http://www.payitforwardfoundation.org/.

Chapter 9: Acting—Step Three

1. See http://www.helpothers.org/cards.php.

Chapter 10: Maturing Into Harmlessness

1. SRI International, "SRI's Values and Lifestyle Program," *In Context* (Summer 1983), http://www.context.org/ICLIB/IC03/SRIVALS.htm.

2. Erik Homburger Erikson, *Adulthood* (New York: W.W. Norton, 1978); *Identity and the Life Cycle* (New York: International Universities Press, 1959).

3. Daniel J. Levinson, *Seasons of a Man's Life* (New York: Knopf, 1978); *Seasons of a Woman's Life* (New York: Knopf, 1996).

4. See overviews of early research in positive adult development in C. Alexander and E. Langer, eds., *Higher Stages of Human Development: Perspectives on Adult Growth* (New York: Oxford University Press, 1990); M.L. Commons, J.D. Sinnott, F.A. Richards, and C. Armon, eds., *Adult Development: Vol. 1. Comparisons and Applications of Adolescent and Adult Development Models* (New York: Praeger, 1989). Information about the Society for Research in Adult Development can be found at http://www.adultdevelopment.org/, along with access to the *Journal of Adult Development*.

5. "Positive psychology" refers to the study of strengths and virtues that help individuals and communities thrive. Further information, research summaries, and questionnaires can be found at http://www.ppc.sas.upenn.edu/, and http://www.authentichappiness.sas.upenn.edu/Default.aspx.

6. Abraham Maslow, "A Theory of Human Motivation," *Psychological Review* 50, no. 4 (1943): 370–396; *Motivation and Personality* (New York: Harper & Row, 1954); *Toward a Psychology of Being* (Princeton: D. Van Nostrand, 1962).

7. D. H. Lajoie and S. I. Shapiro, "Definitions of Transpersonal Psychology: The First Twenty-Three Years," *Journal of Transpersonal Psychology* 24 (1992): 91.

8. See the research presented in Ervin László, *Science and the Re-enchantment of the Cosmos: The Rise of the Integral Vision of Reality* (Rochester, VT: Inner Traditions, 2006).

9. The concept of karma assumes that we have free will to choose to act harmfully or harmlessly and then experience the consequences of our choices. While public attention has focused on the "bad" karma that we experience due to harmful choices, it is also possible for us to accumulate "good" karma due to our positive choices.

10. Michael Price, "Revenge and the People Who Seek It: New Research Offers Insights Into the Dish Best Served Cold," *Monitor on Psychology*, June 2009: 34-37.

11. Michael A. Cremo, *Human Devolution: A Vedic Alternative to Darwin's Theory* (Badger, CA: Torchlight Publishing, 2003).

12. Brian Swimme, *The Hidden Heart of the Cosmos: Humanity and the New Story* (Maryknoll, NY: Orbis Books, 1996), 100.

13. See Paul Bernstein's interview with Michael Washburn, "Life's Three Stages: Infancy, Ego, and Transcendence," http://www.pbernste.tripod.com/life3.htm.

14. See information about the Association for the Advancement of Psychosynthesis at http://www.aap-psychsynthesis.org/. There are now a number of training institutes for psychosynthesis around the world, one of the first of which was the Psychosynthesis and Education Trust, which Assagioli helped to found: http://www.psychosynthesis.edu/.

15. See a copy of the "egg diagram" at http://www.psychosynthesis-uk.com/.

16. Richard Barrett, "What's Right and Wrong with Spirituality in the Workplace" (2008), http://www.valuescenter.com/docs/Whatsright and wrong with spirituality.pdf, 5-6.

17. See discussions in Alice A. Bailey, *Esoteric Psychology, Vol. I* (New York: Lucis Publishing, 1936) and *Esoteric Astrology* (New York: Lucis Publishing, 1951).

18. The Ascendant or Rising Sign refers to the zodiacal sign that is "rising" over the horizon at the moment of your physical birth.

19. Emory University, "Dolphin Cognitive Abilities Raise Ethical Questions, Says Emory Neuroscientist," *ScienceDaily* (February 27, 2010), http://www.sciencedaily.com/releases/2010/02/100218173112.htm.

20. See http://www.culturalcreatives.org/ and Paul H. Ray and Sherry Ruth Anderson, *The Cultural Creatives: How 50 Million People Are Changing the World* (New York: Three Rivers Press, 2000).

21. See http://www.seascale.net/ and Dorothy I. Riddle, *Summary of Research Findings: Spiritual Evolution Assessment Scale™ (SEAS)*, July 21, 2006.

22. Lawrence Kohlberg, *Essays on Moral Development, Vol. II: The Psychology of Moral Development* (San Francisco: Harper & Row, 1984).

23. Carol Gilligan, *In a Different Voice* (Cambridge, MA: Harvard University Press, 1982); Nel Noddings, *Caring: A Feminine Approach to Ethics and Moral Education* (Berkeley: University of California Press, 1984).

24. Dorothy I. Riddle, *Principles of Abundance for the Cosmic Citizen: Enough for Us All, Volume One* (Bloomington, IN: AuthorHouse, 2010), 66.

25. Linda Acredolo and Susan Goodwyn, *Baby Signs: How to Talk With Your Baby Before Your Baby Can Talk* (New York: McGraw-Hill, 2009); see also http://www.babysigns.ca.

26. Suzanne Cook-Greuter, "Mature Ego Development: A Gateway to Ego Transcendence?" *Journal of Adult Development* 7, no. 4 (October 2000): 227-240. See detailed description of language habits in this volume, Chapter 2, page 33.

27. Bailey, *Esoteric Psychology, Vol. I*, 142.

Chapter 11: Our Maturational Opportunity

1. Ervin László, *Science and the Akashic Field* (Rochester, VT: Inner Traditions, 2004), 114.

2. Candace Pert, *Molecules of Emotion: The Science Behind Mind-Body Medicine* (New York: Scribner, 1997), 22-23.

3. Lynn Margulis and Dorion Saga, *Microcosmos* (Berkeley: University of California Press, 1997), 30-31.

4. Alice A. Bailey, *Discipleship in the New Age, Vol. 1* (New York: Lucis Publishing, 1944), 108.

5. Bruce Lipton, *The Biology of Belief* (Santa Rosa, CA: Mountain of Love/Elite Books, 2005), 115-118.

6. Silvia Francesca Maglione, "Effect of Classical Music on the Brain," *Classical Forums*, http://www.classicalforums.com/articles/music_brain.html. See also Don Campbell, *The Mozart Effect: Tapping the Power of Music to Heal the Body, Strengthen the Mind, and Unlock the Creative Spirit* (New York: HarperCollins, 1997).

7. See Amrita Cottrell, *The Encyclopedia of Sound: A Researcher's Guide to Sound and Music in the Healing Arts and Sciences*, http://www.healingmusic.org/, which provides more than 10,000 citations regarding the power of sound and the impact of vibrational energy.

8. McMaster University, "Rough Day at Work? You Won't Feel Like Exercising," *ScienceDaily* (September 25, 2009), http://www.sciencedaily.com/releases/2009/09/090924141749.htm.

9. J.H. Flavell, "Metacognition and Cognitive Monitoring: A New Area of Cognitive-Developmental Inquiry," *American Psychologist* 34, no. 10 (October 1979): 906-911.

10. J.H. Flavell, "Metacognitive Aspects of Problem Solving," in *The Nature of Intelligence*, edited by L.B. Resnick, 231-236 (Hillsdale, NJ: Erlbaum, 1976), 232.

11. Texas A&M University, "Men, Women Give to Charity Differently," *ScienceDaily* (December 28, 2008), http://www.sciencedaily.com/releases/2008/12/081218132142.htm.

12. See Eugene T. Gendlin, *Focusing* (New York: Everest House Publishers, 1978) and also the work of The Focusing Institute at http://www.focusing.org/.

Chapter 12: Creating an Ethic of Harmlessness

1. Parliament of the World's Religions, *Declaration Toward a Global Ethic* (September 4, 1993), http://www.parliamentofreligions.org/_includes/FCKcontent/File/TowardsAGlobalEthic.pdf.

2. Ibid., 1, 2.

3. See information about Rubina Feroze Bhatti, a 2009 PeaceMaker, and the Taangh Wasaib Organization of which she is General Secretary at http://www.taangh.org.pk/.

4. See http://www.avaaz.org/.

5. See the full text of Dr. Abadi's acceptance speech at http://www.mwlusa.org/news/shirin_ebadi_acceptance_speech.htm.

6. See Appendix A for the full text of the UN Universal Declaration of Human Rights.

7. See the text of the 1993 UN Declaration on the Elimination of Violence Against Women at http://www.unhchr.ch/huridocda/huridoca.nsf/(Symbol)/A.RES.48.104.En, and of the 1979 UN Convention on the Elimination of All Forms of Discrimination Against Women at http://www2.ohchr.org/english/law/cedaw.htm.

8. Hans Küng, "Global Ethic and Human Responsibilities," Santa Clara University (April 2005), http://www.scu.edu/ethics/practicing/focusareas/global_ethics/laughlin-lectures/global-ethic-human-responsibility.html.

9. See http://www.goldenruleradical.org/.

10. See http://www.interactioncouncil.org/ and http://www.peace.ca/univdeclarticle.htm.

11. Ibid.

12. Rob Waters, "'War on Obesity Needed for U.S. Health, Milken Says (Update 1)," *BusinessWeek*, http://www.businessweek.com/news/2010-04-26/disease-prevention-key-to-health-care-improvement-milken-says.html.

13. See Johan Galtung's discussion of structural violence in "Violence, Peace, and Peace Research," *Journal of Peace Research* 6, no. 3 (1969): 167-191. Structural violence refers to harm stemming from social structures that oppress or exploit people based on relationships of dominance, such as racism, sexism, classism, ageism, hetereosexism, and ethnocentrism.

14. "Presuppositions" are assumptions of which we are largely unaware because they are buried in every part of our conversations and built into the structure of our language. "Language habits" are our learned language structures or conventions that shape how we view and deal with the world.

15. As part of a Harvard-Peabody study of the Dani, Robert Gardner made a movie, *Dead Birds*, in 1965 that portrayed graphically the war rituals and the manner in which they consumed the lives of the Dani men.

16. Karl G. Heider, *Grand Valley Dani: Peaceful Warriors* (New York: Holt, Rinehart and Winston, 1979).

17. Lest we think that online purchasing of children could not happen in North America, let us remember that, on March 26, 2010, Patrick Molesti of Woodstock, Georgia was captured in Thunder Bay, Canada by the Royal Canadian Mounted Police. Molesti was charged with sexually exploiting children, including trying to buy a five year old boy online, based on evidence on his computer.

18. See http://www.servicegrowth.net/, then Spirituality in Practice/Global initiatives, and http://www.servicegrowth.org/, then Bridging the business-spirituality gap/Global initiatives.

19. See http://www.media-awareness.ca/english/issues/violence/violence_entertainment.cfm.

20. See a detailed discussion of the dynamic of shunning and its impact at http://en.allexperts.com/e/s/sh/shunning.htm.

21. Michelle Ruiz, "Mother Defends Daughter Charged in Bullying Case," *AOL News* (March 31, 2010).

22. See John Wilmerding, "The Theory of Active Peace," http://www.internationalpeaceandconflict.org/forum/topics/the-theory-of-active-peace.

23. See http://www.noetic.org/research/projects.cfm.

24. To download a free copy of the game, go to http://www.curriki.org/xwiki/bin/view/Coll_FJLennon/CoolSchoolWherePeaceRules. For information about the game, see http://www.rtassoc.com/gm_coolschool.html.

25. "Every hour 60 girls are sexually abused, six women undergo clitoral mutilation, and at least one is burned to death or slain by family members in an 'honor killing.' Women are more than six times as likely as men to be the victims of domestic violence, represent over 90 percent of rape victims, and are almost three times as likely to be victims of stalking. Girls comprise the majority of the vast number of children who are forced into prostitution in virtually every country (including North America and Europe) where they are sexually abused 10 to 40 times a day. Almost 70 percent of the casualties in recent armed conflicts have been non-combatant women and girls." From Dorothy I. Riddle, *Principles of Abundance for the Cosmic Citizen* (Bloomington, IN: AuthorHouse, 2010), 115.

26. See http://www.un.org/en/women/endviolence/.

27. See representative Canadian statistics on child abuse at http://www.safekidsbc.ca/statistics.htm and the statistics cited in Appendix B.

28. See full speech at http://www.reliefweb.int/rw/rwb.nsf/db900sid/SMAR-8398KS?OpenDocument&RSS20&RSS20= FS.

29. See statistics cited at http://www.un.org/en/women/end violence/.
30. See statistics at http://www.who.int/mediacentre/factsheets/fs239/en/ and http://www.un.org/womenwatch/.
31. CBC News, "Blog Inciting Hatred Against Women Ruled Legal" (March 31, 2010).
32. Jenny Cuffe, "Child Marriage and Divorce in Yemen," *BBC News* (November 6, 2008), http://news.bbc.co.uk/2/hi/7711554.stm; Alexandra Sandels, "YEMEN: Islamic Lawmaker Decries Child Marriage Ban as Part of 'Western Agenda,'" *Los Angeles Times* (April 18, 2010), http://latimesblogs.latimes.com/babylonbeyond/2010/04/yemen-fierce-opposition-to-child-marriage-ban-persists-among-conservatives.html.
33. See reports at http://www2.ohchr.org/english/law/cedaw.htm.
34. Dorothy L. Sayers, "The Human-Not-Quite-Human," in *Are Women Human?* (Grand Rapids, MI: Wm. B. Eerdmans, 1971), 56-61.
35. See http://hdr.undp.org/en/.
36. FECYT-Spanish Foundation for Science and Technology, "Where Are the Female Scientists in Research Articles?" *ScienceDaily* (December 4, 2009), http://www.sciencedaily.com/releases/ 2009/12/091204092453.htm.
37. See http://www.un.org/en/women/endviolence/about.shtml.
38. See http://www.un.org/en/women/endviolence/network.shtml.
39. See http://www.endvawnow.org/.
40. See http://www.white ribbon.ca/.
41. Caryle Murphy, "A King Takes on Gender Mixing," *The Christian Science Monitor* 102, no. 21 (April 1, 2010), 12. See also http://www.kaust.edu.sa/.
42. Alice A. Bailey, *A Treatise on White Magic* (New York: Lucis Publishing, 1934), 101; 103.

43. For further development of our basic nonduality, see Dorothy I. Riddle, *Moving Beyond Duality: Enough for Us All, Volume Three* (forthcoming, 2011).

44. From a speech given by Dr. Shirin Abadi at the University of San Diego, "Iran Awakening: Human Rights, Women and Islam" (September 7, 2006), http://www.sandiego.edu/peacestudies/ipj/programs/distinguished_lecture_series/biographies/shirin_ebadi.php.

References

Alexander, Charles N., and Ellen Langer, eds. *Higher Stages of Human Development: Perspectives on Adult Growth*. New York: Oxford University Press, 1990.

Anielski, Mark. *The Economics of Happiness: Building Genuine Wealth*. Gabriola Island, BC: New Society Publishers, 2007.

Assagioli, Roberto. *Psychosynthesis: A Collection of Basic Writings*. New York: Viking Press, 1965.

Bailey, Alice A. *Discipleship in the New Age, Vol. 1*. New York: Lucis Publishing, 1944.

———. *Esoteric Astrology*. New York: Lucis Publishing, 1951.

———. *Esoteric Healing*. New York: Lucis Publishing, 1953.

———. *Esoteric Psychology, Vol. I*. New York: Lucis Publishing, 1936.

———. *The Light of the Soul*. New York: Lucis Publishing, 1927.

———. *A Treatise on White Magic*. New York: Lucis Publishing, 1934.

Barlow, Janelle, and Dianna Maul. *Emotional Value: Creating Strong Bonds with Your Customers*. San Francisco: Berrett-Koehler Publishers, 2000.

Barrett, Richard. *Liberating the Corporate Soul: Building a Visionary Organization*. Oxford: Butterworth-Heinmann, 1998.

———. "Origins of the Seven Levels of Consciousness Model." 2005, http://www.valuescenter.com/docs/OriginsSevenLevels.pdf.

———. "The Universal Stages of Evolution." 2010, http://www.valuescenter.com/docs/theuniversalstagesofevolution.pdf.

———, "What's Right and Wrong with Spirituality in the Workplace." 2008, http://www.valuescenter.com/docs/Whats right andwrongwithspirituality.pdf.

Bodhi, Bhikkhu. *The Noble Eightfold Path: Way to the End of Suffering.* Kandy, Sri Lanka: Buddhist Publication Society, 1994.

Campbell, Don. *The Mozart Effect: Tapping the Power of Music to Heal the Body, Strengthen the Mind, and Unlock the Creative Spirit.* New York: HarperCollins, 1997.

Carlzon, Jan. *Moments of Truth.* Cambridge: Ballinger Publishing, 1987.

Chopra, Deepak. *How to Know God: The Soul's Journey into the Mystery of Mysteries.* New York: Harmony Books, 2000.

Commons, Michael L. and F.A. Richards, "Four Postformal Stages." In *Handbook of Adult Development*, edited by J. Demick and C. Andeoletti, 199-219, New York: Kluwer Academic/Plenum 2003.

Commons, Michael L., F.A. Richards, and C. Armon, eds. *Beyond Formal Operations.* New York: Praeger, 1984.

Cook-Greuter, Susanne, "Mature Ego Development: A Gateway to Ego Transcendence?" *Journal of Adult Development* 7, no. 4 (October 2000): 227-240.

Cook-Greuter, Susanne, and M. Miller, eds. *Transcendence and Mature Thought in Adulthood: The Further Reaches of Adult Development.* Lanham, MD: Rowman & Littlefield, 1994.

Cremo, Michael A. *Human Devolution: A Vedic Alternative to Darwin's Theory.* Badger, CA: Torchlight Publishing, 2003.

Dingfelder, Sadie F. "How Artists See." *Monitor on Psychology* 41, no. 2 (February 2010): 40.

———. "Nice by Nature?" *Monitor on Psychology* 58 (September 2009): 60-61.

———. "The Scientist at the Easel." *Monitor on Psychology* 41, no. 2 (February 2010): 34-38.

Economist. "Gendercide: The Worldwide War on Baby Girls." *The Economist* 394, no. 8672 (March 6-12, 2010): 77-80.

Emoto, Masaru. *The Hidden Messages in Water*. New York: Atria Books, 2001.

Erikson, Erik Homburger. *Adulthood*. New York: W.W. Norton, 1978.

———. *Identity and the Life Cycle*. New York: International Universities Press, 1959.

Fischer, K.W. "A Theory of Cognitive Development: The Control and Construction of Hierarchical Skills," *Psychological Review* 87, no. 2 (1980): 477-531.

Fisher, Roger, and William L. Ury. *Getting to Yes: Negotiating Agreement Without Giving In*. New York: Viking/Penguin, 1981.

Fuller, R. Buckminster. *Critical Path*. New York: St. Martin's Press, 1981.

Gendlin, Eugene T. *Focusing*. New York: Everest House Publishers, 1978.

Gilligan, Carol. *In a Different Voice*. Cambridge, MA: Harvard University Press, 1982.

Gladwell, Malcolm. *Blink: The Power of Thinking Without Thinking*. New York: Little, Brown and Company, 2005.

———. *The Tipping Point: How Little Things Can Make a Big Difference*. New York: Little, Brown and Company, 2000.

Goleman, Daniel. *Destructive Emotions: How Can We Overcome Them? A Scientific Dialogue with the Dalai Lama*. New York: Bantam Books, 2003.

———. *Emotional Intelligence: Why It Can Matter More Than IQ*. New York: Bantam Books, 1995.

———. *Social Intelligence: The Revolutionary New Science of Human Relationships*. New York: Bantam Books, 2006.

Hall, Edward T. *Beyond Culture*. Garden City, NJ: Anchor Press, 1976.

_____. *The Silent Language*. Greenwich, CT: Fawcett, 1959.

Hawking, Stephen and Leonard Mlodinow. *A Briefer History of Time*. New York: Bantam Dell, 2005.

Haynes, Carter J. "Holistic Human Development." *Journal of Adult Development* 16, no. 1 (March 2009): 53-60.

Heider, Karl. *Grand Valley Dani: Peaceful Warriors*. New York: Holt, Rinehart and Winston, 1979.

Hochschild, Arlie Russell. *The Managed Heart: The Commercialization of Human Feeling*. Berkeley: University of California Press, 1983.

Hofstede, Geert. *Culture's Consequences: International Differences in Work-Related Values*. Beverly Hills, CA: Sage Publications, 1984.

Hüther, Gerald. *The Compassionate Brain: How Empathy Creates Intelligence*. Boston: Shambhala Publications, 2006.

James, William. *The Principles of Psychology, Vol. 1*. New York: Henry Holt, 1890.

Juergensmeyer, Mark. *Terror in the Mind of God: The Global Rise of Religious Violence*. Berkeley: University of California Press, 2000.

Khamisa, Azim. *From Forgiveness to Fulfillment*. La Jolla, CA: ANK Publishing, 2007.

Kluckhohn, Florence R., and Fred L. Strodtbeck, *Variations in Value Orientations*. Evanston, IL: Row, Peterson, 1961.

Kohlberg, Lawrence. "The Claim to Moral Adequacy of a Highest Stage of Moral Judgment." *Journal of Philosophy* 70 (1973): 630-646.

_____. *Essays on Moral Development, Vol. I: The Philosophy of Moral Development*. San Francisco: Harper & Row, 1981.

_____. *Essays on Moral Development, Vol. II: The Psychology of Moral Development*. San Francisco: Harper & Row, 1984.

Koplowitz, H. "Unitary Consciousness and the Highest Development of Mind: The Relation Between Spiritual Development and Cognitive Development." In *Adult Development, Vol. 2*, edited by Michael L. Commons, C. Armon, L. Kohlberg, F.A. Richards,

Tina Grotzer, and J.D. Sinnott, 105-111. New York: Praeger, 1990.

Korten, David C. *Agenda for a New Economy: From Phantom Wealth to Real Wealth*. San Francisco: Berrett-Koehler, 2009.

Langer, Ellen J. *Mindfulness*. Reading, MA: Addison Wesley, 1989.

László, Ervin. *Science and the Akashic Field*. Rochester, VT: Inner Traditions, 2004.

_____. *Science and the Reenchantment of the Cosmos: The Rise of the Integral Vision of Reality*. Rochester, VT: Inner Traditions, 2006.

Layard, Richard. *Happiness: Lessons from a New Science*. London: Penguin, 2005.

Lehrer, Jonah. *How We Decide*. New York: Houghton Mifflin Harcourt, 2009.

Levinson, Daniel J. *Seasons of a Man's Life*. New York: Knopf, 1978.

_____. *Seasons of a Woman's Life*. New York: Knopf, 1996.

Lipton, Bruce. *The Biology of Belief: Unleashing the Power of Consciousness, Matter, & Miracles*. Santa Rosa, CA: Mountain of Love, 2005.

Loevinger, Jane. *Ego Development: Conceptions and Theories*. San Francisco: Jossey-Bass, 1976.

Lovelock, James. *The Ages of Gaia: A Biography of Our Living Earth*. New York: Norton, 1988.

Lowenstein, George, Scott Rick, and Jonathan D. Cohen. "Neuroeconomics." In *Annual Review of Psychology, Vol. 59*, edited by Susan T Fiske, Daniel L Schacter, and Robert Sternberg, 647-672. PaloAlto: Annual Reviews, 2008.

Mack, Arien, and Irvin Rock. *Inattentional Blindness: An Overview*. Cambridge, MA: MIT Press, 1998.

Margulis, Lynn. *The Symbiotic Planet: A New Look at Evolution*. London: Phoenix, 1998.

Margulis, Lynn and Dorion Sagan. *Microcosmos: Four Billion Years of Microbial Evolution*. Berkeley: University of California Press, 1997.

Maslow, Abraham H. *Motivation and Personality*. New York: Harper & Row, 1954.

_____. "A Theory of Human Motivation," *Psychological Review* 50, no. 4 (1943): 370-396.

_____. *Toward a Psychology of Being*. Princeton: D. Van Nostrand, 1962.

McLaughlin, Corinne, and Gordon Davidson. *Spiritual Politics: Changing the World from the Inside Out*. New York: Ballantine Books, 1994.

McTaggart, Lynne. *The Field: The Quest for the Secret Force of the Universe*. New York: HarperCollins, 2002.

Miller, George A. "The Magical Number Seven, Plus or Minus Two: Some Limits on Our Capacity for Processing Information." *Psychological Review* 63 (1956): 81-97.

Miller, M., and Susanne Cook-Greuter, eds. *Mature Thought and Transcendence in Adulthood: The Further Reaches of Adult Development*. Lanham: Rowman & Littlefield, 1994.

Myers, Isabel Briggs, and Peter B. Myers. *Gifts Differing: Understanding Personality Type*. Mountain View, CA: Davies-Black Publishing, 1980.

Nidich, S., R. Nidich, and C.N. Alexander. "Moral Development and Higher Stages of Consciousness." *Journal of Adult Development* 7, no. 4 (October 2000): 217-225.

Noddings, Nel. *Caring: A Feminine Approach to Ethics and Moral Education*. Berkeley: University of California Press, 1984.

_____. *Women and Evil*. Berkeley: University of California Press, 1989.

Oliver, Richard L. *Satisfaction*. New York: McGraw-Hill, 1997.

Orme-Johnson, D.W. "An Overview of Charles Alexander's Contribution to Psychology: Developing Higher States of Consciousness in the Individual and Society." *Journal of Adult Development* 2000: 199-216.

Peat, F. David. *Synchronicity: The Bridge Between Matter and Mind.* New York: Bantam Books, 1987.

Pert, Candace B. *Molecules of Emotion: The Science Behind Mind-Body Medicine.* New York: Scribner, 1997.

Peterson, Christopher, and Martin E. P. Seligman. *Character Strengths and Virtues.* Oxford: Oxford University Press, 2004.

Piaget, J. *The Construction of Reality in Children.* New York: Basic Books, 1954.

Price, Michael. "Making Sense of Dollars and Cents," *Monitor on Psychology* 39, no. 2 (February 2008): 34-36.

———. "Revenge and the People Who Seek It: New Research Offers Insights Into the Dish Best Served Cold." *Monitor on Psychology,* June 2009: 34-37.

Ray, Paul H., and Sherry Ruth Anderson. *The Cultural Creatives: How 50 Million People Are Changing the World.* New York: Three Rivers Press, 2000.

Riddle, Dorothy I. "Language and Unity." *Sharing,* 1974:10-11.

———. *Moving Beyond Duality: Enough for Us All, Volume Three* (forthcoming).

———. *Principles of Abundance for the Cosmic Citizen: Enough for Us All, Volume One.* Bloomington, IN: AuthorHouse, 2010.

———. "Reclaiming the Principle of Harmlessness." *Esoteric Quarterly* 6, no. 1 (Spring 2010): 17-23.

———. "Spirituality and Politics." *WomanSpirit* 2 (1976): 10-12.

———. *Summary of Research Findings: Spiritual Evolution Assessment Scale™ (SEAS),* July 21, 2006.

———. "The Will in Its Various Forms." *Esoteric Quarterly* 3, no. 2 (2007): 33-36.

———. "Wise Use of Destructive Energy." *Esoteric Quarterly* 4, no. 1 (2008): 25-30.

Roberts, Jane. *The Education of Oversoul Seven*. San Rafael, CA: Amber-Allen Publishing, 1973.

Shields, Stephanie A. *Speaking from the Heart: Gender and the Social Meaning of Emotion*. Cambridge: Cambridge University Press, 2002.

Siegel, Daniel J. *The Mindful Brain: Reflection and Attunement in the Cultivation of Well-Being*. New York: W.W. Norton, 2007.

Simon, Tami, ed. *The Mystery of 2012: Predictions, Prophecies and Possibilities*. Boulder: Sounds True, 2007.

Solomon, Robert C. *The Passions: The Myth and Nature of Human Emotion*. New York: Anchor Press/Doubleday, 1976.

Swimme, Brian. *The Hidden Heart of the Cosmos: Humanity and the New Story*. Maryknoll, NY: Orbis Books, 1996.

United Nations. *UN Universal Declaration of Human Rights*. http://www.un.org/en/documents/udhr/.

Washburn, Michael. *The Ego and the Dynamic Ground: A Transpersonal Theory of Human Development*. Albany, NY: State University of New York Press, 1995.

Whorf, Benjamin L. *Language, Thought and Reality*. New York: Wiley, 1956.

Wilber, Ken. *Integral Spirituality: A Startling New Role for Religion in the Modern and Postmodern World*. Boston: Shambhala Publications, 2007.

Wilber, Ken. "Spirituality and Developmental Lines: Are There Stages?" *Journal of Transpersonal Psychology* 31, no. 1 (1999): 1-10.

Index

Abadi, Shirin, 197, 223
Aboriginal courts, 21
Aboriginal sacred teachings, 124
Absence of Malice, 38
Abstention, 209, 213
abundance, i, 60, 133, 148, 203
action, 60, 61, 62, 137-142, 204, 205-206
adaptability
   Principle of, ii, 36, 169, 174
Adult Learning, 147
Advocacy, 208, 209, 214, 220-222
affirmations, 59
Afghanistan, 198
Ageless Wisdom, 29, 152, 157, 165, 170
ahimsa, 25-26
all-nourishing abyss, 92, 151, 155, 162, 253n3
altruism, 189, 222
ambiguity, 182, 184, 185
American Academy of Pediatrics, 6
anger, 69, 95, 97, 98, 171, 191, 207
appreciative inquiry, 44
Are Women Human?, 218
Armstrong, Karen, 108
artists' perspective, 33
Ascendant, 153, 262n18
Assagioli, Roberto, 152
assumptions, 50, 52, 86, 126, 145, 147, 150, 181
astrology, esoteric, 153, 166
attention, 68-72, 74, 85

   types of, 80-81
attitudes, 32, 50, 60, 145, 192, 215
   Harmlessness Scale™, 208, 235
Authentic Happiness, 111
Avaaz, 197
awareness, 28, 29, 50, 52, 60, 68, 83, 85, 97, 204-205

baby signing, 163
bacteria, 170
Bailey, Alice A., 29, 222
Barrett, Richard, 152, 156, 157
bearing witness, 19, 194-195, 213
behaviorism, 148
beliefs, 32, 50, 145
   limiting, ii, 50-51, 57
Bhatti, Rubina Feroze, 194
Bhutan, 111
Binet, Alfred, 147
birth, 151, 162
blank slate, 154, 162
blindness
   change, 70
   inattentional, 70-71
   perceptual, 71
Blink, 73
boundaries, 154, 155, 164
Brutality, 208, 209, 210, 213, 215-217
bullying, 15-16, 212, 247n19
Burke, Edward, 205
butterfly effect, 58-59
Butterfly Shift , i, 59-63, 75-77, 79, 87, 88, 90, 91, 94, 97, 99, 103, 105,

106, 111, 113, 121, 123, 126-130, 133, 134, 136, 137, 139, 142, 193, 194, 203, 204
  Compassionate, 62, 75, 87, 88, 102, 103, 107, 109, 115-116, 123, 124, 131, 140-141
  Grateful, 62, 87, 88, 102, 103, 107, 109, 116-117, 123, 124, 131, 140, 141
  Joyous, 62, 87, 88, 102, 103, 107, 110, 115, 117-118, 123, 131, 133-134, 141-142
  mini-immersion, i, 60, 61, 67, 83, 85, 91, 95, 97, 112, 113, 118, 131, 135, 204
  Step One, 61, 67, 80, 85-90, 91
  Step Three, 62, 80, 137-142
  Step Two, 62, 80, 113-119

Canada, 19, 22, 218, 221
caring, 121, 135, 161, 195, 208
change, 36, 55, 90, 158, 205
  fear of, 206-208
  lasting, 48, 55-58, 63
  need to, 48, 52-53
  of attitudes, 60
  process, 48, 52
Chanon, Angeles, 212
choice, 2, 67-68, 150, 206
Chomsky, Noam, 26
Chopra, Deepak, 43
Christianity, 25, 28, 40, 107-108
code of conduct, 1, 201
co-evolution, 179
cognitive bias, 79-80
cognitive capture, 71, 86
cognitive tunneling, 71
collaboration, i, ii
Committee on the Eliminations of Discrimination Against Women (CEDAW), 218
compassion, 67, 92, 102, 103, 106, 107-109, 110, 112, 114, 115, 116, 137, 151, 174, 189-191, 195, 203, 208, 213, 223
  Charter of, 108-109, 200
  fatigue, 116
competition, 2, 37, 167
complexity, 73, 80, 146, 174, 182, 183-185, 208
complicity, 16, 195
consciousness, 151
  continuity of, 149, 150, 151, 164
  shift of, 206
contrast, 52
control, 2, 173
Cook-Greuter, Susanne, 33, 152, 164
Cool School, 214
cooperation, 1, 162, 188, 205, 222
  Principle of, ii, 1, 37-38, 161, 174
cosmos, i, ii, 34, 36, 83, 146, 154, 162, 163, 164, 167, 169, 175, 180, 192
Cremo, Michael, 151
crimes against humanity, 7, 10, 11, 12, 16, 17
crisis, 48, 53, 179
Cultural Creatives, 156
cultural filters, 125-130, 147
cultural flexibility, 139
cultural sensitivity, 89, 115, 125, 197
curiosity, 44, 86, 186, 189

Dalai Lama, 1, 42, 107
Darwinian, 161
de Chardin, Teilhard, 167
Dead Birds, 203
decision making, 93-94, 174, 180-182, 186
Declaration Toward a Global Ethic, 193, 198
Defensiveness, 209, 212
degradation, 12, 16, 35, 187
Democratic Republic of Congo, 17

detachment, 134-135, 165, 187, 188
development, 169, 181
    adult, 147, 154, 173
    child, 147
    moral, 161-162
differentiation, 158, 163, 164
dignity, 11, 12, 14, 20, 124, 166, 193, 198
Dismissiveness, 209, 211-212
dissonance, 49
diversity, 7, 108, 186, 193
dolphins, 155
duality, myth of, iii, 53, 223
Dugum Dani, 202-203
Dynamic Ground, 151

Earth, 23, 28, 47, 155, 201. *See also Gaia.*
education, 44
Education of Oversoul Seven, 149
"egg diagram," 152
Elderhostel Institute, 147
elephants, 32, 197
emotional contagion, 103-105
emotional engagement, 60, 204, 205
emotions, 91-112
    and decisions, 106
    as catalysts, 92
    as motivators, 93
    intensity, 94, 100-103
    negative, 95, 97, 98-99, 101, 104, 105
    positive, 95-97, 98, 100, 101, 105, 107, 111, 113, 114, 119
    range, 94, 97-100, 119
    universal, 98
    valence, 94, 95-97
Emoto, Masara, 105
empower, i, 41, 61, 172, 185, 187, 191, 195, 205-206, 223
energetic beings, i, 35, 170, 172, 173, 205

energetic shift, 88
energy, 31, 32, 59, 60, 61, 62, 75, 90, 91, 92, 93, 101, 105, 106, 111, 113, 116, 118, 128, 170, 171, 172, 173, 175, 177, 182, 183, 187, 192, 203, 205
    boundaries, 116
    entanglement, 38, 119
    field, 1, 16, 19, 31, 36, 38, 42, 59, 92, 95, 169, 172, 203, 210
    impact, 100
    waveform, 170, 172
    work, 44
enough, iii, 1, 47, 52, 53, 68, 107, 118, 121, 145, 159, 160, 161, 204, 208, 220, 222, 223
Enough For Us All, i
entitlement, 2, 176, 177
environment, 4
Erikson, Erik, 147
ethics, 150, 173
    global, 193, 199, 200
    of care, 161
    of harmlessness, i, ii, 48, 58, 63, 109, 113, 119, 136, 145, 155, 167, 173, 175, 192, 193-223
    of reciprocity, 26-28, 29, 200-204, 220
European Union, 147
evaluation, 182, 191
expectations, 48, 55, 63, 80, 82, 126, 133, 142, 205, 210
experimentation, 54, 163, 173, 191, 207

fairness, 188
fear, iii, 1, 2, 6, 20, 48, 69, 95, 98, 101, 152, 177, 203, 207, 213, 222, 223
feedback, 36, 61, 130-132, 136-138, 207
feeling, 62, 113-120

feelings, 60, 92, 93, 96, 97, 101, 103, 105, 113
figure-ground, 79, 81, 83, 122, 157, 185, 207
flow, 85, 167
focus, i, 36, 39, 52, 58, 60, 61, 62, 67-83, 85, 86, 87-89, 94, 99, 101, 103, 105, 107, 109, 111, 113-115, 121, 124, 129, 135, 138-139, 141, 146-149, 151-153, 155, 157, 159, 161, 162, 164, 165, 173, 176, 178, 183, 185, 189, 191, 192, 193, 200, 203, 204, 206, 207, 215, 220
focusing, 189
forgiveness, 21-22, 151, 181, 191, 206
From Forgiveness to Fulfillment, 22
frustration, 2
Fuller, R. Buckminster, i

Gaia, 28, 47. *See also* Earth.
galaxies, 163
gender apartheid, 221
gender differences, 92, 95, 101-102, 117, 189, 220
gender discrimination, 198, 218, 220
Gender Empowerment Model (GEM), 220
Gender-Related Development Index (GDI), 220
gendercide, 47
Geneva Conventions, 11, 246n14
Gendlin, Eugene, 189
genocide, 11, 35
  Armenian, 7, 32
Gilligan, Carol, 161-162
Gladwell, Malcolm, 73
Global Campaign for Violence Prevention, 6
Global Virtual Knowledge Centre to End Violence Against Women and Girls, 221

Golden Rule, 200, 202
  Day, 200
  Poster, 200
  Resolution, 200
Goleman, Daniel, 94
goodwill, 29, 136, 174, 187-190, 203
government, 45
gratitude, 62, 67, 102, 103, 109-110, 112, 116, 117, 127, 134, 137, 140, 141, 203, 206
Gross Happiness Index, 111
growth, 39, 170-171, 183
Guess Who's Coming to Dinner, 49

habits, i, 2, 48, 49, 52, 56, 138, 145, 180, 182, 204, 206, 208
Hall, Edward, 126
happiness, 97, 110-111, 148
  psychology of, 39
harassment, 14-16
Harassment, 209, 210, 217-218
harm, i, 1-4, 18-20, 23, 25, 29, 47, 48, 51, 58, 63, 145, 146, 150, 180, 182, 189, 190, 191, 193, 194, 205, 208, 212, 215, 223
  defined, 7-12
  of commission, 12-16
  of omission, 16-18
harmfulness, 2, 4, 6-12, 30, 32, 47, 49, 99, 109, 161, 169, 172, 173, 176, 177, 180, 181, 204, 205
harmlessness, i, ii, 1-2, 4, 18, 19, 23, 25, 26, 29, 34, 45, 47, 48, 49, 58, 61, 90, 121, 131, 133, 145, 154, 155, 156, 161, 169, 171, 172, 173, 174, 176, 177, 180, 181, 187, 191, 192, 193, 203, 205, 208, 222
  as strength, 194-198, 207
  experience of, 48, 63, 90, 136, 142, 205
  gender, 215-222
  habit of, 58, 139, 175, 180, 182, 182, 191-192, 193, 205

in action, 42-45, 222
in thought, 38-40, 222
in word, 40-42, 222
measuring, 208-214
positive, i, 22, 28-29, 47, 63, 203, 212, 223
resistance to, 206-208
Scale™, i, 208, 209, 210, 215, 235
Scale™ credits, 236
Scale™ Questions, 208-209, 235-236, 237-242
Scale™ scoring instructions, 242
Scale™ Scoring Matrix, 235-236, 243
harmonic resonance, 171
health, 44, 74, 202
heart, 106
heart lock-in, 119
Heider, Karl, 203
Hippocratic Oath, 1
Hofstede, Geert, 126
Holocaust, 9
human, 4, 7, 28, 33, 36, 105, 151, 152, 155, 156, 164, 167, 186, 202, 218-219
contact, 53
family, ii, 11, 162, 204, 221
history, 10, 59, 162
potential, 148, 149
rights violation, ii, 6, 8, 19, 215
spirit, 12, 200, 220
Human Development Report, 219
Human Devolution, 151
humanity, 12, 21, 45, 58, 108, 109, 162, 165
humor, 115

immersion, 56-58, 205
harmlessness, i, 58, 60, 61, 63, 85, 141, 142, 145
inattention, 71
independence, 146, 154, 156, 164
individuation, 163

innocuousness, 24-25
Institute of HeartMath, 106, 119
Institute of Noetic Sciences (IONS), 214
Institutes for Learning in Retirement, 147-148
integral theory, 152
integration, 173
intelligence, emotional, 94
intention, 3, 38, 190
InterAction Council, 200
interconnection, i, 7, 36, 205
interconnectivity, 164, 175
Principle of, ii, 1, 4, 7, 29-31, 35, 38, 92, 107, 149, 152, 154, 164, 172, 174, 185, 223
interdependence, i, iii, 29, 108, 109, 154, 155, 161, 165, 174, 193, 202, 205
Principle of, ii, 28, 36, 154, 185, 202
Interfaith Peace-Building Initiative, 200
interference, 171
constructive, 171-172
destructive, 172
International Criminal Court (ICC), 11-12, 17, 19, 247n17
International Day for the Eradication of Violence Against Women, 220
intervention, 121, 198
proactive, 197-198
reactive, 195-196
intuition, 93, 163, 182
Islam, 27, 107, 197, 198

James, William, 68
Journey Is Home, The, 61
joy, i, iii, 39, 67, 83, 92, 98, 101, 102, 103, 110-111, 112, 117, 133, 137, 192, 203, 222, 223
Judaism, 27, 107

282 Enough for Us All: Principles

justice, 7, 11, 20, 108, 150, 161, 166, 198

karma, 42, 149, 150, 165, 261n9
Khamisa, Azim, 22
kindness, 29, 42, 107
   random acts of, 59, 133, 206
King Abdullah University for Science and Technology (KAUST), 221
Kluckhohn, Florence, 126
Kohlberg, Lawrence, 161
Korten, David, 61
Küng, Hans, 199

Langer, Ellen, 73
language, 32, 34, 40-41, 50, 56, 200, 204, 211
   and newborns, 50, 163-164
language habit, 33, 34, 41, 164, 202, 265n14
laughter yoga, 113
Lehrer, Jonah, 106
Levinson, Daniel, 147
life forms, 36, 155, 162, 186, 191, 202
life purpose, 110, 149, 165, 166-167, 192
lifelong learning, 147
   Academy of, 148
linearity, 53, 149, 160
Lipton, Bruce, 171
Locke, John, 154
Lorenz, Edward, 58

Mack, Arien, 70
Margulis, Lynn, 37, 170
Marino, Lori, 155
Maslow, Abraham, 123, 148, 157
maturation, 28, 146, 169, 171, 172, 183, 192
   dimensions, 174-175, 192
   goal, 173, 193

moral, 162
physical, 162, 163, 164, 173
psychological, 63 , 145, 146
task, 165
maturity, 30, 145, 155, 203
media, 44-45
Merton, Thomas, 108
metacognition, 181
metaphysics, 42, 91, 110, 149, 153, 222
microbes, 37
Miller, George, 68
mindfulness, 73-75, 82-83, 86
mindlessness, 73, 194
Morton, Nelle, 61
motivation, iii, 3, 123, 124, 203
Moving Beyond Duality, iii
Mozart Effect, 171-172
multi-dimensional, 96
multi-directional, 171, 173
multi-potentiality, 38
muscle
   emotional, 94-103, 111, 113, 119, 205
   harmlessness, i, 203-204
music, classical, 171
musical analogies, 103, 105, 114, 158, 159, 160
Myers-Briggs Type Indicator, 73, 165

Native American Great Law of Peace, 29
needs hierarchy, 123, 138, 148, 156
Network for Grateful Living, 117
networking, 1, 37
Neugarten, Bernice, 147
New Guinea, 202
Nobel Peace Prize, 197
Noddings, Nel, 161
nonduality
   Principle of, ii, 34-35, 174, 183, 223

nonlinearity
  Principle of, ii, 34, 59, 145, 149, 156, 169, 174, 181
Nonviolence, 209, 214
noticing, 61, 67, 71, 76, 81, 85-90, 91, 113, 121, 142
Nuremberg
  Principles, 9-11, 16, 195
  Trials, 7, 9, 246n13
nurturance, 174, 185-187

Obama, Barack, 43
objective, 146, 154, 155
One Life, 163, 164, 165, 171, 223
optimism, learned, 111

Pakistan, 194-195
paralanguage, 94, 125
Parliament of the World's Religions, 193
participation, 45, 193, 216, 220
  Principle of, ii, 31-34, 68, 69, 154, 174, 177, 179
pathology, 148
pay it forward, 133-134, 142, 206
Pay It Forward, 133
peace, 11, 34, 206, 223
  active, 214
  Department of, 45
  Science of, 214
perception, 51, 68, 71, 126, 154, 176, 183
perspective, 51, 63, 83, 85, 86, 124, 128, 145, 146, 149, 156, 157, 160, 161, 167, 170, 183, 185, 187, 188, 189, 202
Pert, Candace, 92, 105, 170
physics
  Newtonian, 2, 92, 146
  quantum, 2, 38
  wave, 173
Piaget, Jean, 147
Pillay, Navi, 215

Plutchik, Robert, 98
positive adult development, 147
Positive Harmlessness in Practice, i, 236
possibilities, 57, 59, 61-62, 67, 73, 96, 99, 111, 135, 139, 142, 148, 163, 167, 171, 210
presuppositions, 32, 145, 146, 202, 265n14
Prince, Phoebe, 212, 247n19
principles, seven, ii, 1, 13, 29-38, 146, 154, 160, 162, 167, 169, 174, 177
Principles of Abundance, ii, 1, 2, 29, 154, 160, 162, 169, 205
psychology
  gestalt, 73, 79
  humanistic, 148
  positive, 39, 96, 147, 260n5
  transpersonal, 148, 152, 160
Psychosynthesis, 152

rapport, 121-123
"rational man," 146
Rays, 42, 153, 166, 249n16
reality, 31, 154
reframe, 111, 118
reincarnation, 149
religion, 44, 193, 199, 204
resources, scarce, 2
respect, 29, 41, 52, 60, 62, 86, 88, 108, 124, 125, 127, 135, 139, 142, 166, 186, 188, 193, 198, 200, 201, 206, 223
responsibility, 1, 3, 36, 38, 44, 61, 104, 126, 129-130, 146, 150, 154, 156, 161, 172, 174, 177-180, 193, 205, 212
revenge, 18, 150, 188
Rising Sign, 153, 262n18
Roberts, Jane, 149
Rock, Irvin, 70
Rome Statute, 11-13

Rwanda, 17-18, 246n13
Ryan, Catherine, 133
Saudi Arabia, 198, 221-222
Sayers, Dorothy, 218
scarcity, i, iii, 2, 133, 177
  myth of, 1, 207
self-actualization, 123, 138, 148, 149, 152, 167
self-awareness, 165
self-discipline, 53, 174, 175-177, 186, 194
self-identity, 154
self-interest, 1, 28, 29, 53, 156, 157, 162, 187, 189, 207
self-realization, 152
Seligman, Martin, 39, 111
separativeness, illusion of, 154, 156
service, 157, 165
service quality, 122
silence, 41-42
simplicity, 34, 185, 192
skills, 157, 159, 160, 166
Smile cards, 142
Solomon, Robert, 92
soul, 41, 148, 149, 151, 152, 153, 157, 162, 164, 165, 166, 171, 222
sound, 40
Spiritual Evolution Assessment Scale, 160
spirituality, i, 149, 151, 152, 156, 157, 160, 164, 165, 167
sports analogies, 55, 157
stages, developmental, 147, 156-161, 162, 167, 169, 192
Stanton, Elizabeth Cady, 177
stereotypes, 50-51, 95, 211
  gender, 51, 56, 92
strengths, 25, 148, 166, 192
Strodtbeck, Fred, 126
Supporting Spiritual Development, 7, 206
Supportiveness, 209, 213
survival of the fittest, 2

Sweden, 147
Swimme, Brian, 151, 162
synchronicity, 172

Tanzania, 197
Tariq Khamisa Foundation, 22
TED, 108
tension, 2, 25, 30-31, 36, 91, 92, 154, 198, 210
thin-slicing, 73, 86
thoughts, 23, 38-39, 92, 105, 194, 222
timing, 136
tipping point, 48, 193
trust, 38, 98, 11, 182
Truth and Reconciliation, 19, 34

unique, 29, 50, 110, 111, 149, 154, 162, 163, 164, 165, 166
UNiTE, 6, 220-221
United Kingdom, 197
United Nations (UN), 6, 7, 215, 218, 221
UN Convention on the Elimination of All Forms of Discrimination Against Women, 218, 247n16
UN Declaration on the Elimination of Violence Against Women, 198
UN Development Fund for Women (UNIFEM), 217, 221
UN Development Programme (UNDP), 219
UN Millennium Development Goals, 7
UN Network of Men Leaders, 221
UN Universal Declaration of Human Rights, 11, 124, 198, 200, 204, 246n16
United Religions Initiative, 7, 44
United States, 19, 20, 43, 202, 247n19

Universal Declaration of Human Responsibilities, 200-201
unitive, 149, 152
unity, 7, 149, 162
universe, iii, 1, 163, 169, 173
"us-them," 3, 35, 154, 183

values, 32, 38, 49, 52, 53, 124m 125, 145, 146, 173, 176, 180, 182, 197, 199
  continuum, 126
  universal, 124-125, 173
Values & Lifestyle Program, 145
Values-Conscious Business, 206
vibration, 169, 170, 173
violence, 1, 2, 4-6, 11, 12, 17, 18, 20, 22, 23, 25, 26, 34, 35, 47, 51, 58, 59, 60, 61, 63, 108, 145, 154, 164, 169, 183, 187, 195, 197, 198, 202, 204, 206, 207, 208, 223, 245n3
  structural, 202, 265n13
visualization, 91, 116, 258n3

wan'ni, 195
Wan'ni: Murdered Marriages, 195
Washburn, Michael, 151
water, 105, 170

well being, 165, 166-167, 177, 191
White Ribbon Campaign, 221
Whorf, Benjamin, 32, 56
Wilber, Ken, 152, 156
willful blindness, 195
Wilmerding, John, 214
win-lose, 2
win-win, 186, 206, 213
women, 19, 51, 56, 108, 197, 202, 203, 215-220
  violence against, ii, 6, 19-20, 32, 58, 195, 197, 198, 203, 210, 212, 215-220, 221, 223, 266n25
Women's International War Crimes Tribunal, 19-20
World Health Organization (WHO), 4, 6, 7, 217
worldview, 2, 32, 33, 79, 125, 145, 146, 164, 204, 205, 214
  Literacy, 214
  Transformation, 214

Yemen, 218
"young-old," 147
Youth Courts, 20

Zambia, 197

List of Exercises

Motivation Regarding Harm	3
Violence in Media	5
Addressing Harm	8
Identifying Harmful Actions	13
Identifying Harmful Inactions	18
Forgiveness — Part 1	21
Forgiveness — Part 2	22
Redefining Innocuousness	24
Behaving Harmlessly	28
Interconnectivity and Harmlessness	31
Participation and Harmlessness	32
Nonlinearity and Harmlessness	34
Nonduality and Harmlessness	35
Interdependence and Harmlessness	36
Adaptability and Harmlessness	37
Cooperation and Harmlessness	37
Harmlessness in Thought	39
Harmlessness in Word	40
Harmlessness in Action	43
Recognizing the Need for Change	49
Identifying Cues for Change	53
Motivation to Change	54
Change Process	57
Absorption	69
Shifting Attention	70
Focused Attention	72
Mindfulness	74
Being Mindful	75
Who We Notice	77
Noticing Helpfulness	78
Optimism and Pessimism	82

Choosing Your Focus   89

Making Decisions   93
Momentum of Emotions   96
Positive and Negative Emotions   99
More Intense Positive Emotions   100
Emotional Contagion   104

Becoming Noticed   122
Cultural Flexibility   130
Giving Feedback   132
Initiating Change   134
Detachment   135

Assumptions About Development   148
Developmental Context   150
Our Starting Point   153
Relating to Others   155
Developmental Process   158
Identifying One's Life Purpose   166

Examining Values   174
Evaluate Yourself on Self-Discipline   175
Enhancing Self-Discipline   176
Evaluate Yourself on Responsibility   178
Enhancing Responsibility   179
Evaluate Yourself on Decision Making   180
Enhancing Decision Making   181
Evaluate Yourself on Complexity   183
Enhancing Complexity   184
Evaluate Yourself on Nurturance   186
Enhancing Nurturance   187
Evaluate Yourself on Goodwill   188
Enhancing Goodwill   189
Evaluate Yourself on Compassion   190
Enhancing Compassion   191

Proactive Harmlessness   196
Harmlessness and the Golden Rule   199
Challenges of Harmlessness   207
Dealing with Dismissiveness   211
Nonviolence Without Harm to Self   214